Choreographies of Resistance

Geopolitical Bodies, Material Worlds

This series publishes studies that originate in a range of different fields that are nonetheless linked through their common foundation: a belief that the macro-scale of geopolitics is composed of trans-local relations between bodies and materials that are only understandable through empirical examination of those relations. It is the interaction of these elements that produces the forces that shape global politics, often with outcomes that differ from the predictions of macro-scaled theories. This world poses questions: how do materialities such as the built environment and the body reproduce global power structures, how are they caught up in violent transformations and how do they become sites of resistance? How do assemblages of human and non-human elements both fortify and transform political space? What possibilities for political change are latent within the present?

Series Editors

Jason Dittmer, Professor of Political Geography at University College London
Ian Klinke, Associate Professor in Human Geography and Tutorial Fellow at St John's College, Oxford

Titles in the Series

The Geopolitics of Real Estate: Reconfiguring Property, Capital and Rights, Dallas Rogers
Choreographies of Resistance: Mobile Bodies and Relational Politics, Tarja Väyrynen, Eeva Puumala, Samu Pehkonen, Anitta Kynsilehto and Tiina Vaittinen.

Forthcoming

Moving Lives: Critical Reflections on Climate Change and Migration, Andrew Baldwin and Giovanni Bettin
Territory Beyond Terra, Kimberley Peters, Philip Steinberg, and Elaine Stratford
The Politics of Bodies at Risk: The Human in the Body, Maria Boikova Struble

Choreographies of Resistance

Mobile Bodies and Relational Politics

Tarja Väyrynen, Eeva Puumala,
Samu Pehkonen, Anitta Kynsilehto
and Tiina Vaittinen

ROWMAN &
LITTLEFIELD
INTERNATIONAL
London • New York

Published by Rowman & Littlefield International Ltd
Unit A, Whitacre Mews, 26-34 Stannary Street, London SE11 4AB
www.rowmaninternational.com

Rowman & Littlefield International Ltd.is an affiliate of Rowman & Littlefield
4501 Forbes Boulevard, Suite 200, Lanham, Maryland 20706, USA
With additional offices in Boulder, New York, Toronto (Canada), and Plymouth (UK)
www.rowman.com

British Library Cataloguing in Publication Data
A catalogue record for this book is available from the British Library

ISBN: HB 978-1-7834-8672-4
PB 978-1-7834-8673-1

Library of Congress Cataloging-in-Publication Data Available
ISBN: 978-1-78348-672-4 (cloth : alk. paper)
ISBN: 978-1-78348-673-1 (pbk. : alk. paper)
ISBN: 978-1-78348-674-8 (electronic)

∞™ The paper used in this publication meets the minimum requirements of American National Standard for Information Sciences—Permanence of Paper for Printed Library Materials, ANSI/NISO Z39.48-1992.

Printed in the United States of America

Contents

Chapter 1

Introduction

Choreography, Mobility and Politics

> All those who had waited on the dark Moroccan beach had been at the mercy of human smugglers. [...] They had been loaded into the ship's hold like modern day slaves. They were not shackled with iron chains but they were chained by their dreams, despair and fears that have made them leave, and seek freedom in Europe. [...] She had drifted in cold water and hit a rock. [...] On a cold Spanish shore searchlights had blinded her and dogs had sniffed their way to her. The border-guards had looked at her with tired eyes. (Mankell 2001, 6–7)

Henning Mankell's novel *Tea-Bag* (2001) provides a powerful introduction to some of the themes that are also central to this book. In the opening scene of his novel, Mankell describes the experience of migration from the perspective of a mobile subject who travels from Africa to Europe. Mankell's novel addresses events that have become prominent since the eighties, with the gradual imposition of the Schengen visa system that was introduced to control and halt human mobility into Europe (e.g., Bigo and Guild 2005). Owing to the lack of legal routes, the organization of travel is largely operated by networks of human smugglers. Over the years, migration across the Mediterranean has cropped up every now and again in public discourse and has been defined as a 'crisis' at different entry points between the Gibraltar Strait, the island of Lampedusa and the Aegean Sea between Greece and Turkey, not to forget the crisis of mass arrivals to the Canary islands in 2006. The latest of these 'crises' began in 2015 with the arrival of more than a million migrants and refugees to Europe over one year. However, not everyone made it to European soil: over 3,770 deaths were documented as people drowned in their quest to flee persecution or acquire better, more worthwhile lives for themselves.

In 2015 mobility became, once again, subject to the most intense political debate within the European Union (EU). The member states struggled both nationally and as a bloc to find ways by which they could govern and effectively manage the growing number of arriving asylum seekers. These events failed to produce any fundamental move towards opening legal access across the EU's external borders but, on the contrary, strengthening external border controls became once again the more prominent feature of the debate. New ad hoc internal border control measures were introduced at various European borders, calling into question the very existence of the Schengen Agreement that has guaranteed free movement within most of the EU area. These measures aim to curb the 'uncontrolled' arrival and circulation of migrating bodies that are regarded as undesirable (cf. Kynsilehto 2014). At the same time, demographic changes within the EU countries have led some to conclude that controlled and selective immigration is one way of addressing the changes as there is need for skilled workers within the EU.

Back to Mankell's prognostic novel. The reader of the novel soon realizes that the migrant whose experience is presented in the scene is called Tea-Bag. Through Tea-Bag and her body that is constantly on the move, Mankell describes a relational body that moves in space, and by so doing enacts its own agentive capacity. In the scene cited at the beginning of this chapter, Tea-Bag's body crosses a geographical and political border with great difficulties: the trip with human smugglers is dangerous, and she is forced to fight for her life in the cold sea. The novel depicts border control practices embodied by armed border guards and their dogs that find her too late to ensure her return to Morocco. The border described in the novel represents more than a mere geographical and political marker. It is enacted materially, and although it spatially separates Africa from Europe, it is fluid in its functions (Mountz and Loyd 2014).

The struggle that is depicted in Mankell's novel unfolds also in reality at the borders of contemporary Europe. It illustrates the ways in which practices of governance intertwine with practices of resistance that mobile subjects enact. As in contemporary Europe, Tea-Bag's corporeal and lived experience of the border and boundaries is filled with darkness, coldness, fear and struggle. The stories of crossing waters form a core in many of the migrants' narratives not only in Mankell's book, but also in the lives of those migrants that struggle to reach not only European territory but also other shores around the globe (e.g., Perera 2013). Migrants row across rivers, lakes and seas; travel in small fishing boats or inflatable rubber boats; and are picked up by border guards' fast boats and military ships. During these physically demanding journeys migrants invest their bodies at full, risking life and limb to enter Europe and to cross the continent in order to reach a place where life is envisaged as liveable. The mechanisms of control never fully capture the migrant's body in this dangerous attempt to reconfigure the division between those who are allowed to enter and move about, and those who should stay away—a division which is ultimately about those who belong and those who don't.

Mankell's novel also illustrates the power that individual stories can have in creating understanding about how exclusions and inclusions are materially enacted and how the practices of control are undermined. We engage with this dynamic in this book. We draw on empirical descriptions of different choreographies of resistance and their interacting entanglements, and discuss mobility as a material-corporeal phenomenon. The book focuses on mobile bodies and their capacity for obstructive and resistant action by which the bodies disrupt existing scripts and orders that others seek to impose on them. In so doing, we reflect on how positionalities and agentive capacities are created and regulated in various places and spaces. Through developing the notion 'choreographies of resistance', we shed light on the ways in which resistance and governance are enacted in mundane and daily relations that yet are connected—in some cases more explicitly than in others—with networks of power and political control. Instead of addressing the body in constraining and abstract terms, we focus on the concrete fractional and mundane choreographies of material bodies that refuse to become firmly and unambiguously located in the systems of power (Edkins and Pin-Fat 2005; Kynsilehto and Puumala 2015; Puumala 2013, 2016; Rygiel 2011, 2012; Vaittinen 2014). We illustrate how political agency manifests itself corporeally, with multiple consequences and challenges to the disciplinary practices that regulate bodies. We show how the mobile body is more than a mere target of territorial politics that aims to situate bodies within the state.

Illustratively, Mankell's novel *Tea-Bag* depicts how crossing a national border does not guarantee freedom. Rather, it takes the main character to the liminal space of the refugee camp. The camp itself is a de-territorialized space surrounded by fences that prevent her from entering into Europe. The camp is located in Spain, and yet it exists nowhere. No country wants to accommodate the inhabitants of the camp, since they represent the 'dangerous aliens' to be managed and kept away from the core of the (nation)state whose existence is seen to be dependent on a coherent national identity. Tea-Bag is not designated to the category of migrant worker, but positioned as one of those who do not belong. As the story of Tea-Bag evolves, the borders she experiences and that her body crosses become even more fluid and ambiguous in their effects. The borders and their effects are navigated and resisted, which reveals that the mobile body is not completely powerless in the face of governmental practices.

Tea-Bag ultimately reaches Sweden, where her existence becomes characterized by attempts to hide and remain invisible to a variety of surveillance practices, including surveillance monitors and CCTVs, digital traces of money transactions and ID checks. However, her solitary life in Sweden is filled temporarily with a sense of community and response-ability among other migrants with whom she leads a life that is largely invisible from the state's perspective. Despite being invisible to the structures that characterize modern societies, Tea-Bag's is a life based on resourceful and creative strategies of survival that include living in empty apartments, using stolen mobile phones and acquiring false IDs. Tea-Bag's body resides in the national space and within its systems of governance and, at the same time, she does not exist for the state.

Through attending to different forms of mobility, such as asylum seekers, undocumented migrants, migrating beggars and migrant nurses, and also the empirical cases included in the book, we position ourselves to attest to the scope and diversity of corporeal and resistance in the realm of politics. These case studies bring forth a variety of resistances and forms of agency that enable us to conceive politics as a relation and allow discussion of the kinds of material-corporeal consequences and effects that emerge through resistance. The focus enables us to conceive subject positions in terms of relational positionalities that take shape and gain their meanings in events as people come together or as they are subjected to governmental practices. Our focus on choreographies of resistance enables us to fathom power in terms of networks and variously 'demarcated axes' (Kynsilehto 2011a) that become operationalized in different contexts through multiple practices, instead of claiming that governance creates fixed positions and permanently marginalizes some and privileges others.

In the course of our book, we show that *deeply disadvantaged and vulnerable bodies that are not assumed to have any agentive capacity, let alone political agency, can obstruct and resist the smooth functioning of practices of governance, and thus open a space of political resistance and action.* In order to grasp the agency and resistance of the mobile bodies we seek to 'reveal the *agency* of the seemingly excluded' (Ackerly and True 2006, 249, emphasis in original). For us, agency is about the abilities and the potential of the material and relational body to enact political choreographies anew (Puumala and Pehkonen 2010; Puumala, Väyrynen, Kynsilehto and Pehkonen 2011; Vaittinen 2014), not so much about subject positions. Consequently, acts of resistance may take place in mundane encounters, and they often disrupt the functioning of the practices of governance targeted at mobile bodies. Political agencies discussed in this book concern the boundaries built between bodies and the efforts to cross these boundaries, so as to raise a different political relationality. Although the kind of doing, corporeal and gestural body politic that we study might sometimes appear as being without clear purpose and direction, it illustrates that the struggles and the everyday relations of mobile bodies evoke a variety of reconfigurations, capacities and potentialities that allow for political agency.

As a whole, Mankell's topical novel suggests three interrelated features of the (inter)national order of things which are of our interest in this book. We perceive the national order of things as a dominant categorical order of modernity (Malkki 1995a, 1995b), through which bodies, spaces and politics are governed within and in-between states. First is the *territoriality* of the nation state, which portrays nation states as inescapably territorialized entities, and mystically rooted in particular, geographically delimited sites (Rajaram 2013; Mountz and Loyd 2014). Thus, with the national community imagined as inseparable from its designated place in the world, the national order of things contains, within itself, a tendency to emplace and territorialize whichever categories it attempts to organize: bodies in particular. Secondly, while *nationality* is no more given that any other feature in our socially constructed identities, it nevertheless remains at the core of one's political identification in the modern world. Thus, so as to be a right-bearing subject in a national order of things, one must belong to a nation with its designated territory. Thirdly, *belonging to a nation is not only a matter of 'being part of' or belonging to a larger whole, but a territorialized relation that has corporeal implications on individuals, and their mobility*. Namely, just as the idea of nation is 'rooted' in a particular territory, so are the bodies that represent different nations inherently 'incarcerated, or confined, in those places' (Appadurai, quoted in Malkki 1992, 29; emphasis in the original). This at least is the discursive order of the (inter)national order of things, by which the world works. However, as this book as well as many other accounts in the field of mobility studies testify, by virtue of their mobility, mobile bodies continue to challenge this territorial order of rootedness which, after all involves many other actors, discourses and material practices than those of the nation state. In other words, the state as a fixed entity is also vulnerable to movement.

Mankell ends his novel with a scene where Tea-Bag reminisces the act of crossing the border that separates Africa from Europe and envisions her future:

> A bridge was in front of us and the last part of the trip was like a leap into a vacuum. [...] I just know that the bridge that we thought we had seen will be constructed. [...] One day the pile of bodies lying on the sea bed will be so high and the bridge made out of skulls and ribs will form such an embankment that no border guard, no dog, no drunken sailor, no human smuggler will be able to tear it down. (Mankell 2001, 377–378)

Mankell's horrific, yet insightful, imaginary of an embankment made of corpses evokes a strong sense of materiality. In the scene, the flow of human bodies is so immense that its material force will challenge nation states' capacity to monopolize the space of the political and determine who can be considered politically active and have agentive capacity. Mankell's novel suggests that the sheer number and force of corporeal movement will contest the control and governance of migrants' bodies. This is the imagery that became prominent in the summer and autumn of 2015, with massive media attention to hordes of people traversing countries on foot and crammed into trains and buses. In this book, however, we are more interested in the minutiae and mundane choreographies that engage bodies in resistant action than in the 'flow of migrants'. Through our empirical research we show how the human body is always capable of changing its prescribed choreography, no matter how strictly people are governed and how completely their spaces of agency and movement are limited. Here, we draw on the work of feminist scholars, critical geographers and new materialists. This wide range of theoretical traditions enables us to understand the body as simultaneously a material 'fact' of life and as a realm of discursive governmentality. In other words, the body is 'multiple' (Mol 2002): it is created in and through various simultaneous (material) practices, enactments and configurations, and it can be a myriad thing and be in different places at the same time. Agency is, in this view, about visibility and response-ability (cf. Barad 2007), about the possibilities of becoming visible and of mutual response, which is not to deny, but to attend to the power and practices of governance. In so doing, the book contributes to the study of the political significance of movement, mobility and the nonverbal as the resistant body poses a challenge that is both praxical and philosophical: it ultimately invites us to reconsider political space and fathom the meanings and content of relational politics.

MOBILITY

Although the number of people on the move in the world today is indeed historically high, migration is not a new phenomenon. Mobility is a defining characteristic of human life. People have always moved from one place to another, in search of food and livelihood, because of love, work, conflicts and disasters. These migratory movements may be temporary or seasonal, or they may be of more permanent nature. They may be about very short physical distances, or trajectories that reach across the globe. Sometimes the movement is forced, other times it is voluntary. Very often it combines elements of both, thus becoming something in-between these two. While it is difficult to pinpoint an exact, universal let alone ahistorical definition of human mobility, there is one persistent and unavoidable feature that is always present, namely, the relational and material body that moves in space, while making space for its own (political) existence—indeed, while making space political.

The burgeoning interest in issues related to human mobility within what has been named 'mobility turn' (Sheller and Urry 2006) in social sciences has produced important intersections of different academic fields. Alongside the traditional fields of inquiry—human geography and social anthropology—political scientists have demonstrated an increasing interest in migration and mobility studies (see, e.g., Salter 2007; Squire 2009, 2011, 2015). Over the past two decades or so, there has been an increasing scholarly interest, for example, in the ways that border control measures organize individuals on the move into categories of desired and undesired travellers (e.g., Salter 2006; Adey 2009; Amoore and Hall 2009; Fassin 2011; Bigo, Carrera and Guild 2013; Kynsilehto 2014; Häkli 2015). For those considered to be desirable entrants to the national territory, border controls have been increasingly lifted, such as within the European space of free movement, the Schengen area. As for the travellers categorized as undesirable, ethnic profiling and other forms of creating figures of suspicion enhance the use of internal border controls that indeed extend the border to basically any area, sometimes far from the geographical borders (e.g., Fassin 2011; Mezzadra and Neilson 2013). These control measures have become increasingly corporeal ranging from fingerprinting to full-body scanning (Amoore and Hall 2009). Consequently, not all those willing and often forced to leave their countries are entitled to do so, at least in the sense of choosing their destination. In this vein, human mobility can be considered a resource that is differentially accessed (e.g., Cresswell 2010).

As indicated above, this book discusses mobility within the territory of the EU. The EU today is one of the most glaring manifestations of globally uneven mobilities as noted earlier. Since the Maastricht Treaty and its further developments, especially the Tampere European Council of 1998, the EU has designed a common policy on migration and asylum. Since the 2010s, this European migration regime seems to have been developing in two different directions, with labour migration differentiated from that based on humanitarian and family reasons. In addition, along with the successive enlargements of the EU to cover more of Central and Eastern Europe, new mobile populations have emerged within the Schengen area. One of these groups is the Roma people from Central and South-Eastern Europe. In many receiving states, their mobility across the internal borders of the EU, and particularly their presence on the streets as buskers and beggars is represented as a problem. Nevertheless, as EU citizens, the Roma people have the same rights to intra-EU mobility as any other ethnic group, and the resistance towards their visibility on the streets thus tends to unfold as racist complaints directed at legislation. In chapter 5 we provide a detailed empirical account on how these racialized corporeal choreographies are enacted as well as resisted in corporeally entangled events of relational encounters.

As for the mobility of labour into Europe, despite the economic downturn, in the political rhetoric at least, migrants considered as skilled or highly skilled are said to be welcome, especially in fields of work where the national pool of labour is lacking, such as the health sector. Despite the rhetoric, however, and apart from the most highly skilled labour being eligible for the Blue Card, such people are known in EU jargon as 'the best migrant talent', institutional racism persists in the field of labour migration. Consequently, entry to professional life remains differentially accessible for non-EU citizens, particularly if coming from the global South (e.g., Raghuram 2004; Fannin 2010; Kynsilehto 2011 a&b). The European Qualifications Framework (EQF), while making educational credentials comparable among EU citizens, tends to discriminate against the credentials of those from outside the EU. Consequently, skilled professionals from outside the EU often end up working in lower-skilled jobs than their training allows for. This is particularly so in health care and the service professions, where the lowest paid and 'dirtiest' jobs in Europe have over time developed into 'migrants' jobs', with non-EU doctors working as nurses, registered nurses as caregivers, engineers as cleaners, teachers as domestic workers, etc. (e.g., Kynsilehto 2011 a&b; Vaittinen, Sakilayan-Latvala and Vartiainen forthcoming; Kingma 2005; Cuban 2013). These racialized choreographies of mobility will be discussed in chapter 3, through the case of Filipino nurse migrants who are classified as the global corporeal reserve of care.

However, within European policies and practices as well as in public discourse, those with a work visa are made more welcome than those who migrate for family reasons, seek asylum, or those who otherwise have limited or no access at all to a visa or residence permit. Mapped as humanitarian, 'irregular' or 'undocumented' migrants, their presence is seen as unwanted, a problem to be fought against; yet, the state actors are increasingly reluctant to find ways for these persons to legalize their stay. In the meantime, while waiting for some solution to their situation, people on the move reside in various kinds of camps across the European territory (see Valluy 2005). These vary from reception and detention facilities (Puumala 2016), formally established and operated to more or less temporary refugee camps resembling the established refugee camps sited elsewhere around the globe, to the makeshift dwellings that have emerged in the vicinity of border areas (Laacher 2002; Rygiel 2011; Agier 2013; Lafazani 2013; Kynsilehto 2014). As has become widely visible over the course of 2015 and later, these forms of residence are not only a quasi-permanent feature of particular localities such as Calais and its surrounding area in northern France, but also sites of manifold contestations and corporeal politics. Since mid-November 2015 the successive closing of borders across the so-called Balkan route resulted in the village of Idomeni on the Greek side of the border between Greece and the Former

Yugoslav Republic of Macedonia (FYROM) becoming one of the most exposed new hubs. First the borders along the Western Balkans were closed to people of all nationalities except Syrians, Iraqis and Afghans; gradually they were closed to everyone. As a result, some ten thousand people have been blocked in Idomeni in a makeshift camp on the fields and railroad tracks. On several occasions, people blocked at the border have demonstrated against these border practices and the critical living conditions in the campsite. The border guards on the Macedonian side of the border have responded by firing tear gas and rubber bullets randomly into the crowd that includes many children and people with severe disabilities. These incidents illustrate the crude war that is being waged against bodies deemed as undesirable. In chapter 4, we will discuss corporeal politics in the context of camps, combining insights from makeshift camps and squatted houses in Calais with those from reception facilities for asylum seekers in Finland.

As illustrated above, humanitarian migrants are concretely represented as a security threat and treated as such. Moreover, they are represented as a 'burden' to the welfare state, with additional policy measures implemented across Europe that are legitimized by the claim that they decrease this 'burden'. In Denmark, for example, it is now legal to confiscate asylum seekers' assets (over a given limit) before accommodating them in camps, and similar policies are in places also in other European countries (IRIN News 2016). The critiques of these policies have rightly pointed to a chilling historical comparison with the Third Reich, to the processes where the Jews were first labelled and then gradually reduced to lesser human beings, and where eventually they had their belongings confiscated prior to their dispatch to the camps and the gas chambers.

Indeed, asylum is attested to as being in crisis. On the one hand, the critics defending human rights have, for several years, pointed to the current asylum regime and the provisions of the Geneva Convention of 1951, as too limited in their definition of 'refugee', which despite the existing complementary frameworks for subsidiary or humanitarian protection pushes many of those who cross borders in search of a safe haven into a position of 'undocumented' migrant, when the law fails to recognize them as refugees. On the other hand, particularly since 2015, racist voices that include not only populist politicians and extreme right groups wishing to close the European borders from asylum seekers but also more mainstream political leaders and heads of states have called for the reconsideration of the values of protection as spelled out in the Geneva Convention. Many hold the view that, with the present '*Völkerwanderung*' across the Mediterranean, Europe is at the limits of its 'capacity' to provide space for those on the move. This calls, they argue, if not the rewriting of international law, at least its re-reading and suspension. In the most concrete move thus far to externalize border control and, hence, access to asylum, on 18 March 2016 the EU concluded a deal with Turkey that permits fast returns of people who arrive on the Aegean islands. This deal became operational immediately. Since the first returns undertaken within this new framework, it has been disclosed that the human rights guarantees, which were superimposed on the deal following multiple critiques by international human rights watchdogs, were not respected in practice (Kingsley 2016).

ON CHOREOGRAPHY

Despite the attempts at the national and international level to, first, control the movement of bodies and, second, to produce categories of desired and undesired bodies in relation to their access to the European and national space, the body on the move has agentive and political capacity that intertwines these practices of governance. Allegedly marginalized human bodies are capable of refusal, disturbance and resistance. As indicated in Mankell's novel, the body's agentive capacity is multiple, and even contradictory. In the spirit of Tea-Bag's story, it will be illustrated in this book that a body's agentive capacity and political agency emerge in situational encounters with others and in spaces beyond any single subject's capacity to govern. Hence, we contribute to the literature that engages with the political potential of bodies resisting governance, and we seek to do that in ways that do not merely imply the intentionality of the agency. This requires us to move from the verbal to the embodied terrain of agency and action. Through the three case studies presented in the book, we show how the body, and its myriad ways of being, often go beyond the linguistic frameworks of communication as refusal and resistance often belong to the non-verbal domain (also Puumala et al. 2011). In our view, many of the existing theories on mobility and agency fail to engage with the full political potential of bodies on the move, namely with their capacity to resist and obstruct the smooth function of governance and reconfigure the hierarchies of those who belong to the community and those who do not. In order to contribute to the literature on mobility, governance and resistance, we turn to the idea of choreography—defined here as *a situationally enacted interaction framework that provides both sequential and governing structure for bodily events* (cf. Pehkonen 2016)—for several reasons that we discuss next.

First, focusing only on the ways in which mobile bodies are governed is problematic to the extent that, ironically, it ultimately fails to engage with movement itself. As Erin Manning (2007, 132) posits, movement is 'intrinsic to the relations between bodies and worlds'. Movement, in our view, is a foundational category of thinking, so in order to fully appreciate the corporeal element of mobility, we prefer the term 'choreography'. The term 'movement' is often understood as something solipsistic, topographic, mechanical and linear, whereas the notion of choreography implies non-linear embodied relationality, extending and reaching out towards others. The empirical examples given in the book show how especially the stereotypical understanding of the 'unwanted mobility' and its implications on political agency, is limited and misleading. We show how these bodies have more agentive capacity and political potential than usually assumed when they are viewed only as objects of either government, containment, surveillance or bureaucratic practice.

Second, choreography is a force of affective potential as it articulates a body's capacity to communicate and integrate with other bodies. In this sense, choreography is always social and connecting. It is based on forms and trajectories of movement that leave a visible yet impermanent trace in interaction and identities. Sometimes, the trace is invisible and more permanent in nature. This is particularly in the case of traumas that challenge the dichotomic orders between body and mind as well as the linear temporal orders of bodies in space (Väyrynen 2016). The tired eyes of the border guard that meet and 'welcome' the castaway Tea-Bag to Europe in Mankell's novel, for instance, indicate that the encounter is primarily not between two individuals connected in space through the process of life-saving rescue. Instead, the choreographies of border crossing reduce the status of this event within networks of power to that of a routine task: for the guard, Tea-Bag is just another 'body' procedurally handled and maybe later on reported for statistical purposes; for Tea-Bag, the guard may be her rescuer but only in order to prepare her for just another ambiguous leap into a state of limbo. Through choreographies, we examine the oscillation of bodies as lived, experienced and material configurations, which are simultaneously extremely personal, shared and relational. The body is always partially marked and 'already-made-visible', yet simultaneously always in the process of becoming, with other bodies in particular, historically contingent choreographies.

Third, choreography is a physical and conceptual catalyst that negotiates between risk and control in the production of space and the exploration of physical relationships. Because of its common association with dance, choreography is often understood primarily to signify composing and arranging dances in advance. Etymologically, however, the ancient Greek word *choreo* derives from *chora*, which since Plato has been used to denote the earliest western spatial thinking (Kymäläinen and Lehtinen 2010). Choreo means the being in, passing, entering to or holding space. Choreography can thus be interpreted to refer to the writing of practices of being (corporeally) in space and inhabiting space. In choreography motion fills space between places, just like in interaction the reach of a hand fills the space between the person reaching and the person being reached out to. In addition, this spatial motion becomes named through interaction: a hand is reaching out in order to take something away, to give something or to gesture denial (see Puumala and Pehkonen 2010). In this manner, the body exposes fixed places as spaces with no firm meaning (cf. Logan 2005). The movement between places (from where one left and where one imagines arriving), in choreographic terms, is never exhaustively prescribed, making it possible for Tea-Bag to 'enact the political' (Rancière 1998), exhibit political agency. Through the three case studies of this book, we show how spaces for political agency never simply pre-exist, but are (re)articulated through bodies' choreographies (cf. Manning 2007 & 2009). This takes place in their always-already emergent intra-action (Barad 2007) with other bodies, which continues to (re) configure the political space in its becoming.

Fourth, authors on interaction analysis have used the notion of choreography to highlight the situational nature of identities and action. Although often loosely defined, the concept is related to analysing social action as well as movement and change, and thus the approach aims at bracketing any prefix structures of action or identities as the primary way of explaining and understanding social life (e.g., Aronsson 1998). It emphasizes dialectical epistemology in social space where talk, identity and context are co-constitutive. Instead of fixed identities, choreographies are playgrounds for gaining 'hybrid subjectivities'. Katrina Brown (2012, 802) writes about 'becoming hybrid in particular ways, the myriad forms of body-to-body attunement demanded, and related struggles over who bears responsibility for orchestrating coexistence, and how that shapes the kinds of public spaces produced'. Choreography therefore offers a non-conventional way of thinking about the emergent and constantly reconfigured entanglements of bodies, places, spaces and identities as well as of matter and discourse. Choreographies are situationally enacted in *events* (see also chapter 2). Thus, it is crucial to emphasize, the primary object of the study in this book and in its three case studies is not just to present abstract or generalizable instances of bodies named as asylum seekers, labour migrants, undocumented migrants or Roma people. Rather, in elaborating the choreographies of corporeal mobility, we focus on *events* that emerge in a variety of corporeal-material encounters.

Finally, in addition to choreography cutting across various conceptual discussions, it is deployed in our book also as an analytical and methodological tool that oscillates between ethnographic inquiry and analyses of social interaction. The book is generally about choreographies of mobility, which describes more than the techniques of governing the physical movement and classifications of border-crossing bodies. Choreography is not only a script by which people move on the stage, or in the pre-inscribed political space. Instead, choreographies are enacted in the corporeal practices that people deploy, in order to form and maintain the movement within these governmentally written scripts. These practices are partly made of interactional techniques, mundane members' methods (Garfinkel 1967) as well as of the actors' absorbing the disciplinary and governing rules. For the actors, choreographies are thus always partially pre-given, already planned and expected to be treated as fixed lines to be followed—yet, the interactional resources of bodies can be used to remould the situation. Looking closer at these bodily techniques of interaction reveals that choreographies are in fact open to surprises, and even disturbances, and tend to produce extraordinary acts out of the ordinary (Laurier 2004). The choreography of one kind can thus be exploited and turned into some other type of choreography: a queue into a rush for safety, a modest question into an accusation in a conversation. Sometimes the event itself 'demands attunement to, and sometimes concession of, differential hybrid subjectivity, since it is not always possible for all desired subjectivities to be performed at once' (Brown 2012, 807).

To summarize, our aim is to think about bodies' expressive extending that creates skinscapes that 'are as complex as the landscapes that make up our political vistas' (Manning 2007, 112). This ontological understanding of the body as material-political-relational leads to a more complex approach towards the relationship between a body, body's agentive capacity and the political. Our approach questions the notion of the individual subject in the first place and shows the ways in which relations between bodies and the world unfold, how the body is a site and a space of 'the political' and how resistance emerges from the body.

CHOREOGRAPHY AND THE STUDY OF POLITICS AND MOBILITY

Through its case studies, the book illustrates events where different types of mobile bodies, from asylum seekers, undocumented or labour migrants to beggars enact political spaces through embodied encounters and interaction with other bodies, both mobile and immobile, human and non-human. The choreographies we describe are not readily generalizable to apply to all asylum seekers, labour migrants and beggars in Europe at all times. Yet, they are configured in the context of governance that is politicized, planned and motivated by the desire of effective state control of boundaries and clear-cut classification of mobile bodies in the national order of things—that is, in a liminal context, where migrants in Europe and beyond are asked to improvise space for their embodied existence.

A juxtaposition of choreographed and improvised events is often found in literature on choreography (e.g., Tulbert and Goodwin 2011, 87). Logical as it might sound, the relation is not all that straightforward, however, since improvisation works only as far as the participants can relate the improvised acts to the prevailing action. People cannot improvise randomly: the acts of improvisation are picked up from a reservoir of commensurable acts within certain situations. This has relevance also for the possibilities of resistance. When found on the shores of Europe by the dogs and border guards, Tea-Bag did not overtly resist the arrest. Had she tried, the next sequences of the ongoing choreography would only have included moves and techniques intrinsic to the roles and the tasks of the border guards. Resistance, as outlined in this book, is played out in moves and actions that make the prevailing script limp and ineffective as well as make something which was not visible earlier visible, even if only partially or momentarily so. Tea-Bag's resistance takes the form of an undercover agent, an identity partly based on deception, on resourceful and creative survival strategies. She is not visible to the nation, yet she leads a resourceful life.

It cannot be too strongly emphasized that our perception of choreography goes beyond governance as a predesigned script. To be effective, the scripts of governance need to be recurrently enacted by interacting bodies in space. This also means that, while research often tends to cast its gaze on the bodies of the migrant only, the political choreographies of human mobility we want to examine are never only about 'the migrant'. In order for the choreographies of mobility to become enacted, there are always other bodies invited to participate in this improvised, spatial and creative material and corporeal 'installation' that reproduces and reconfigures the national order of things, in minuscule local events just as much as at the level of high politics and global governance of borders. In the case studies, the readers of this book will thus encounter (and interact with) not only bodies of 'the migrant', but also a variety of other bodies: from the researching bodies to elderly bodies in need of migrant care to a man who films beggars on the street to a sex toy called Barbara doll. In encountering and interacting with these choreographies that are not performed by the migrant bodies alone, perhaps the readers will participate in their reconfiguration, too.

Corporeal choreographies do not occur in fixed frameworks, but they reconfigure the existing frameworks. In adopting such a stance, we choose to read the mobile body as a possible agent that can open routes to imagine connections between bodies, forms of being-in-common and reaching out towards other bodies (cf. Matteo 2005). As further elaborated in chapter 2, in our examination of the corporeal choreographies of mobility, we acknowledge the fact that political agency and resistance, which (may) result in the body moving itself beyond national body politics, are not always easily recognizable, visible or articulable. Indeed, as Steve Pile (1997, 3, cited in

Takhar 2007, 133) posits, they can operate subtly in spaces, not defined by those in authority and therewith more difficult to control. If we accept the claim that becoming is an essential part of being and that bodies do not reside in space, but space in bodies, it is meaningful to explore whether and how the body can function as a site of political resistance and as a space for political action (also Puumala et al. 2011; Manning 2007, 58; Vaittinen 2014). In this vein of thought, choreography becomes also a question of methodology.

CHOREOGRAPHY AS AN ANALYTICAL AND METHODOLOGICAL APPROACH

The central elements of choreography—space, movement and relationality—make mobility an ideal site of study in an effort to cast light on how bodily choreographies work. As the twists and turns of the travels of Tea-Bag indicate, the moving body would be extremely difficult to approach with a prefixed and static analytical framework. Movement being characterized by constant becoming and being-in-between, it is difficult to designate one analytical and methodological tool for the study of the relational and material body that moves in space, while making space political. To capture these entangled corporeal processes, we propose again the use of the notion of choreography, now as an analytical and methodological approach that recognizes and respects the ontological eventness of corporeal choreographies.

When examining the movement of bodies across time spaces, one effectively examines the trajectories of mobile bodies in space and time. In mobility studies, the term 'migrant trajectory' is often used as an analytical tool. As Tiina Vaittinen (2014, 196–197) elaborates, according to a dictionary definition, the term 'trajectory' refers to the 'path' of a body 'moving under the action of given forces', for instance, 'a curve of a projectile in its flight through the air'. When it comes to the movement of relational and material human bodies across borders, however, the mere description of relevant forces at any given moment is an extremely complex task, which can only ever succeed partially, through situated knowledges (cf. Haraway 1988). After all, in human mobility, the trajectory is not just about a physical object moving (or being moved by external forces) from one place to another: human mobility is always a lived and embodied phenomenon (Kynsilehto 2011a, 1547), defined by an indefinite number of material and relational factors, such as entangled corporeal configurations of gender, ethnicity and class. Thus, 'migrant trajectories are rarely, if ever simple. What is consistent in them is the almost constant movement' (Puumala 2016, 8). Consequently, when studying migrant trajectories, 'the researcher necessarily halts the movement at the moment of her or his observation', meaning that 'the migrant trajectory is always an image captioned by the researcher for specific purposes', that is, 'a multifaceted vehicle of corporeal relatedness extracted from a space of complexity' (Vaittinen 2014, 197).

However, studying choreographies of mobile bodies is not only about the analysis of singular migrant trajectories. Studying corporeal choreographies is about analysing the intersections and entanglements of migrant trajectories with/in a discursive-material world, and how these entanglements become reconfigured in particular events. Here, it is important to note that the migrant body, its subject position or its mobile trajectory in space is not the force of resistance *per se*. Instead, the resistance of the mobile body emerges from its capacity to weave together innumerable material-discursive networks of relatedness, when navigating through the global space (Vaittinen 2014, 197). The choreographies of resistance examined in this book are thus about criss-crossing movement of the body and between bodies that resist and mould pre-existing scripts of governance. Here, the spaces of agency do not 'simply pre-exist, but are articulated through bodies' movements' (Puumala et al. 2011, 86).

Thus, given that corporeal choreographies of resistance become into being in events, studying choreographies of mobile bodies requires empirical approaches that defy strictly predetermined methodological, theoretical and conceptual frames. In this book, event-based, situational choreographies have required us to work across several research fields and methods (see Collins 2008), with the methodologies, conceptual and theoretical frames of each chapter fitted to and emerging with the particular empirical phenomena under scrutiny. In this regard, as indicated earlier, the case studies are perhaps best understood as research apparatuses, in the Baradian sense (Barad 2007). Building on Niels Bohr's quantum physics, Karen Barad (2007, 143; our emphasis) portrays apparatuses as 'macroscopic *material arrangements* through which particular concepts are given definition, to the exclusion of others, and through which particular phenomena with particular determinate physical properties are produced'. These material arrangements, she emphasizes, are not simply about preconfigured laboratory setups; they involve the discursive arrangements and theoretical concepts, just as much as the measuring equipment, including the researching body and her/his particular gendered, classed and racialized entanglements with the phenomenon under scrutiny. For social scientists used to acknowledging the researcher's embeddedness in the world she or he studies, this is a familiar story of course. However, to comprehend this book's line of argument, it is crucial to recognize each chapter as emerging from particular research apparatuses, where particular material-discursive arrangements, methods and concepts uncover particular corporeal choreographies of resistance.

Research apparatuses are always design(at)ed to measure, make visible and comprehensible certain phenomena rather than others, thereby simultaneously producing the particular phenomena under observation (rather than others). The case studies of this book ranging from Filipino migrant nurses in the postcolonial global political economy and Finnish care of the elderly to the embodied resistance of asylum seekers and undocumented migrants in different types of camps in Finland and France to the Roma travellers on the streets of Vantaa, the use of disparate research apparatuses has been a methodological necessity. To the extent that analysing corporeal choreographies is about the minutiae analysis of embodied, material-relational events, no single overarching methodology can ever be assigned to the study of choreographies, but each methodological apparatus needs to emerge as part of the phenomena under scrutiny. This explains, for instance, why we utilize a variety of complementary theoretical conceptions of the body and resistance in different chapters of the book: each empirical case requires its particular conceptions, and corporeal choreographies, evented as they are, cannot be analysed by fitting empirical cases to a single pregiven rigid theoretical frame.

Furthermore, to the extent that the choreographies emerge as part of the book and its writing, the chapters should not be read as mere theoretical reflections on a pre-existing phenomenon. The book, in itself, is choreography of a kind, where particular analyses of particular choreographies of resistance are weaved together in a wider apparatus of mobility and embodied eventness of resistance. Barad (2007, 146) argues that 'apparatuses are themselves phenomena', 'constituted and dynamically reconstituted as part of the ongoing intra-activity of the world', and this book and the research herein are part of these choreographies of becoming. Finally, the apparatuses/phenomena of the three empirical case studies are also a wider global apparatus/phenomena of corporeal choreographies, which we will weave together in the concluding chapter.

In the subsequent chapter, we will elaborate the conceptual and theoretical dimensions of our disparate apparatuses, with a reference to previous literature. As for methods utilized in the case studies, three empirical chapters build on ethnography (chapters 3 and 4) and interaction analysis (chapter 5). First, so as to work towards getting a grasp of the inherent relationality of corporeal being on the move, we have resorted to ethnography, which in this book is not to be understood as denoting simply field-based methods of gathering research material. Rather, we work from the premise that ethnography is a methodology that entails the research process as a whole, beginning from the theoretical ideas with which the researcher engages with his or her field in participant observation, observing participation and engaged listening. Like David Harvey writes about the body:

> [the body does not] float freely in some ether of culture, discourses, and representations, however important these may be in materializations of the body. The study of the body has to be grounded in an understanding of real spatio-temporal relations between material practices, representations, imaginaries, institutions, social relations, and the prevailing structures of political-economic power. [. . .] It is a nexus through which the possibilities for emancipatory politics can be approached. (Harvey 2000, 130; see also Barad 2012)

Multi-sited ethnography (Marcus 1995; 1998) is often seen to be best suited for the study of bodies on the move as it moves out from single sites and 'fields' and relies on a variety of research material. In ethnographic research analysing the kinds of materialities, discourses and spatio-temporal relations to which Harvey referred in the above quote, is taken seriously. The same applies to the corporeal presence of the researcher at different phases of the research process. As a corporeal being herself or himself, the researcher always moves with the research apparatus, while simultaneously entangled with other bodies, and it is through her or his material, relational body that the insights are filtered as a specific reality. However, ethnography has not been very popular among political geographers, political scientists or IR scholars (though, especially within mobility studies; see, e.g., Doty 2001, 2010; Penttinen 2008; Squire 2014, 2015). However, in an effort to 're-people political geography' (Megoran 2006), political geographers have used, for example, participant observation in their attempts to build up a fuller understanding of mobile bodies and international boundaries (e.g., Mountz 2004, 2010; Agier 2008, 2013; Hiemstra 2014) and their own emotional involvement in the field (e.g., Woon 2013). Feminist scholarship committed to ethnography of the state has suggested ethnography as a way to think 'through the potentially aporetic quality of [...] state in practice' (Coleman and Stuesse 2016; also Williams and Massaro 2013). Through a rich variety of ethnographic data, chapters 3 and 4 will discuss how states and governmentality are enacted and how the enactments of power are always connected with multiple events of resistance.

In interaction research in turn, which forms the central methodological tool in chapter 5, the emphasis is typically on data analysis rather than in the processes of acquiring the data. The appropriate data consists of audiovisual recordings of actual events as these recordings allow the researcher to look at the event as it unfolds. The repeatability of the actual event makes it possible to concentrate on the micro-level interactional units, gestures and moves that would otherwise go unnoticed even by the participants. The data can take the form of recordings specifically produced for research purposes, but this is not always available in the cases of violent events or institutional settings for example, and documentary films or third-party recordings provide an equally useful source. Chapter 5 on street begging in this book elaborates this type of methodological approach, being based on a detailed analysis of videos posted on YouTube. These 'badly produced' videos were selected 'for good analytical reasons' (Laurier 2015) to study the choreographies and outcomes of resisting begging, events situated in the Finnish urban space.

To sum up, choreography for us is not just a theoretical concept, or a theory of mobility and resistance, but it is a methodological approach that requires making visible—for disparate audiences—the events, hidden situations and resistances that bubble under the unforeseen sea of corporeal relationality. It is also an ethical approach, to the extent that it suggests that we are already entangled in the very same choreographies of mobility as bodies perceived as 'others'. These mundane entanglements should make the book readable, as a space of unexpected encounters, for those who know nothing about those 'other' among 'us', who think we have nothing in common, or who opine that the refugees, beggars and migrant workers have nothing to do with our lives. Indeed, choreography does not allow us to explain particular phenomena and events with fixed categorizations or predetermined identities whichever they are. It also challenges our understanding of the limits and forms of resistance as it entices us to engage with interaction and events that situationally disrupt learnt ways of acting and thinking. By focusing on the body, choreography makes us see resistance in such events and forms that are often omitted. Learning to live these choreographies and remaking them as inclusive for those who make us what we are the ultimate suggestions choreographic analysis makes.

A NOTE ON RESEARCH MATERIAL

Although this book is a product of collaborative writing, the research material for the case studies has been collected individually. Chapter 3 is based on Tiina Vaittinen's multi-sited ethnography of the global care economy, conducted in ageing Finland between 2011 and 2015, and in the Philippines in June 2014. The research material consists of a great variety of ethnographic research material including periods of participant observation among a group of recently graduated Filipinos, life-story interviews and informal chats, participant observation at Filipino social gatherings in Finland, interviews and non-taped discussions with Finnish employers, recruiters, state and labour union officials, a taped interview at the embassy of the Philippines in Helsinki as well as focus-group interviews at governmental offices in the Philippines and informal and formal discussions with Filipino academics and nurses and nurse advocacy groups in the Philippines.

Chapter 4 draws on material collected in two multi-sited ethnographic research projects. Eeva Puumala has collected data in four different reception centres and a detention unit in Southern Finland. The empirical insights of chapter 4 are based on interviews, discussions and field notes with asylum seekers who were either still waiting for a decision concerning their asylum claim or had recently received the decision—most often a negative one. Anitta Kynsilehto's data from Calais draws on her multi-sited ethnography that maps different 'hubs' of transit migration in Europe (France, Greece, Italy and Spain) and in the Southern and Eastern Mediterranean (Algeria, Morocco, Tunisia and Turkey). Her project is conducted with undocumented migrants on their way to a country where some form of settlement might be possible. In these trajectories being on the move or in transit may last for years. This material is combined with insights from solidarity actors who engage in diverse ways of working with undocumented migrants. The main bulk of data from Calais and its surrounding area was collected in parallel with an inter-associative field mission, as a result of which a critical report denouncing different forms of human rights violations in the Northern Littoral was published (EMHRN 2011). In chapter 4, additional data from journalistic texts is used.

The YouTube videos analysed in chapter 5 are part of a data corpus on public displays of hate speech and aggressive talk, collected by Samu Pehkonen. The aim has been to show and study in detail the verbal and non-verbal means participants use in real-life public events to express their emotions, feelings of injustice or to degrade their conversational partners. The videos have been transcribed using notation system developed in conversation analysis and multimodal interaction analysis. To contextualize the events displayed on the videos, official documents, news articles and blog texts on Roma beggars have been used.

Chapter 2

Exploring Choreographies of Resistance

Returning to the main theme of this book, namely, mobile bodies and their capacity for obstructive and resistant action by which the bodies disrupt existing scripts and orders, we need to discuss resistance more fully. In chapter 1, we introduced the notion of 'choreographies of resistance' and geared the attention tentatively to resistance and governance that are enacted in mundane and daily relations and that are connected with networks of power and political control. The question of resistance as it becomes manifested by the material body in the context of mobility is a complex one. Through the empirical cases of the book and research apparatuses that uncover particular corporeal choreographies of resistance, we will attest to the scope and diversity of the corporeal as it brings forth a variety of resistances and forms of agency that enable conceiving politics as a relation and discussing what kind of material-corporeal consequences and effects emerge through resistance. We will illustrate how resistance is played out in moves and actions that make visible something which was not visible earlier, even if only partially or momentarily so. In order to grasp the complexity we seek to understand the following questions in this book. What becomes of resistance when it takes place in mundane encounters and often disrupts the functioning of the practices of governance targeted at mobile bodies? What kind of political positionalities emerge when differently categorized bodies come together with institutional structures and practices, or with other bodies? What are the processes through which spaces are accommodated for resistance? Ultimately, we are interested in this book how resistance(s) happen in events.

In this chapter, we continue to create a conceptual-analytical framework from which we derive in the empirical cases. First, we will introduce our take on resistance and argue for the necessity to understand resistance as a multifaceted phenomenon, which is not only collective, intentional and connected to wider social mobilizations, but at the same time is also evented, situational and relational. We argue that if we are to understand those choreographies of resistance that mobile bodies enact and their political significance, both of these dimensions of resistance need to be taken into account and connected with the wider political, societal, economic and material contexts within which they take place. Secondly, we will discuss the connections of our approach to practices of governance and the various networks of power that become operationalized in the context of human mobility. This is done to better understand how mobile bodies are subjected to various practices of governance that are designed to narrow the space of agency and delineate bodies in particular—although by no means unambiguous and non-productive—ways with regard to both the state and the prevailing postcolonial order of global capitalism. Besides constraining the space of political agency, labels and practices of categorization also open a space of tension between inclusion and exclusion, governing and being governed, control and resistance (Puumala and Pehkonen 2010). Concentrating only on the voice created or made possible by the status and location in the hierarchies of the national order of things would mean remaining within a logic that counts on predefined identities. We claim that even though political agency and resistance are often conceived as being created through situations and statuses imposed on people, it is imperative, as Shaminder Takhar (2007, 123) notes, to take into account the body's capacity 'to make social change, to resist structural constraints, and to challenge racist and sexist discourses'.

This brings us to the final part of this chapter where we address the body and bodily agency from the perspective of new materialism and ponder what kind of political positions or conditions for political existence can be fathomed through that framework. In this, we take a cue from the idea of an ontological, material body as interacting, expanding, reaching out. We show how political agency can be understood to unfold as not only multilayered but also infused with contradictions, ambiguities and ambivalences. What we aim to imply is that instead of a single way of acting politically, there are shifting modes of agency and enacting in ways that bring to the fore the political. As a whole, this chapter paves the way for an understanding of the kinds of possibilities for resistance that emerge when we approach the evented existence of the material body and take into account its capacity for obstructive behaviour and its inherent creativity and resourcefulness.

MULTIFACETED RESISTANCE

Resistance, within our framework, denotes a material practice or event that communicates people's understanding of their surroundings and negotiation of their capacities, or reorganizes relations that exist between people. In this interpretation of resistance, we come close to Diana Coole and Samantha Frost's (2010, 10) idea of 'choreographies of becoming' in which subjectivities are 'constituted as open series of capacities or potencies that emerge hazardously and ambiguously within a multitude of organic and social processes'. In this book, we focus more on those political, economic and social processes that pertain to the constitution of the mobile body and its positionalities. We also touch upon the organic processes of the body organism in passing at the end of chapter 3, while discussing the functions of oxytocin at the caring touch of a dying body. For us, resistance, or more correctly resistances do not follow a certain strategy, but rather there are 'multiple mobile and transitory points of resistance' (Foucault 1990). Hence our focus on events and on the interaction in which mobile bodies participate. Given this approach, there is a need to move away from associating political identities with obvious forms of resistance, and to move towards the multiplicity of experiences and power (cf. Takhar 2007), as people on the move make subtle political articulations and claim their bodies in the face of practices of governance and control.

Over the past three decades, there has been a scholarly shift among critical geographers from interest in the uncontested dynamics of hegemonic space and powerful agents to studies of resistance that aim to 'uncover the different ways in which seemingly stable representations of space [are] opened and changed' (Rose 2002, 383). For example, Paul Routledge (1993) has elaborated the social movements approach in efforts to challenge state-centred notions of hegemony and power and contest the colonization of the 'political' by the state. In a similar vein, Tim Cresswell (1996) has examined acts and individuals that appeared to transgress hegemonic socio-spatial boundaries and expectations about spatial behaviour. Furthermore, the book that Michael Keith and Steve Pile (1997) edited on geographies of resistance exemplifies the turn from sites of oppression and discrimination to spaces of resistance, while bringing the radical view of postcolonial and queer theory to spatial politics. These resistance studies in geography are based on an assumption that 'resistance is a conscious practice intentionally designed to overcome or change some perceived effect of power' (Rose 2002, 385) while unintentional resistance—conceptualized as 'transgressions' by Cresswell (1996, 23)—'encompasses acts that have subversive and potentially emancipatory effects but which are not conceptualized in terms of conscious ideological struggle'.

Critical voices have emerged too. The turn from hegemony to resistance has later been addressed in terms of 'romance of resistance', where resistant agency is seen in the non-spatial and non-temporal terms of 'some seminal and heroical universalized human spirit' (Sparke 2008, 423). According to Matthew Sparke, this type of approach preempts empirical research as there seems to be no need for recognizing the spatial and temporal contexts and intersections of resistance. Another critical point, raised by Mitch Rose, concerns the nature of the social system in resistance studies which begins with—and remains captured by—power. By relying on a stable centre, resistance studies cultivate hegemony rather than question the very presence of systems (Rose 2002).

In the study of politics and international relations, a major turn in the study of resistance took place thirty years ago, when the Marxist and Gramscian perspectives on resistance moved away from class-based ontology and started to emphasize social movements. The movements were seen as operating within civil society, being instrumental and intentional in character, oriented strategically towards material goals and aimed at an enemy, such as the financial elite or the state (see, e.g., Colás 2002). The scholarship established a clear distinction between the politics of governance and the politics of resistance, where resistant social movements and mobilization are seen to create a non-authoritarian, participatory political culture as well as strengthen participatory democratic practice and significantly alter political life. Those more influenced by post-structuralism or constructivism tend to be interested in the latter aspect, namely, how movements re-inscribe political practices (e.g.,Walker 1994; Bleiker 2005), or how movements can be understood as new forms of political agency united around marginalized identities (Campbell 1998; Bleiker 2000; Polet 2007; Shaw 2003).

In research geared towards understanding human mobilities, resistance has figured in an array of ways following similar lines to the developments outlined above. Mobility as such has been analysed as resistance, with resistance conceived as acts of rebellion against authorities and hegemonic cultural ideals that are not straightforward (Cresswell 1993). Migration as resistance is also a recurrent theme: the act of leaving one's country as a protest to the prevailing order (Coyle 2007), and the modality within which this move is enacted as another form of migration as resistance (Killias 2010). Many social movements have emerged from and/or accompanied migrant struggles as a part of a wider quest for social justice (Zaman and Tubajon 2001; Nyers 2006). These movements often centre around the right to make one's claim for asylum heard and recognized, and to obtain a legitimate migration status. Migrants without a residence permit and, hence, without an acknowledged status in a national order of things (Malkki 1995b), make claims for a voice and a recognition of their presence in order to rewrite the contours of this very order. These claims not only address the established order of allocating residence permits and statuses with accompanying allocation of bodies to their assigned places, however, but they also bring forth the internal dynamics and political contestations within and between solidarity networks (Rygiel 2011; Millner 2011; Squire and Darling 2013; Squire 2015).

Our approach in this book is towards the unintentional 'haecceities' (Garfinkel 1967), the indexical character of any situation where resistance is always-already part of social relations. Hence the focus is neither on 'transgressions' nor on participatory political culture. Rather, our aim is to develop a reading of resistance that would be able to account for the multiple and ambiguous ways in which the world is remade and engaged with, without neglecting the role that the networks and enactments of power play. Within such a framework, practices and events of resistance take form situationally, within particular space-time, a 'simultaneity' 'of a multiplicity of trajectories', accounting Doreen Massey's (2000, 225) conception of space. Our argument is that a nuanced understanding of choreographies of resistance is needed, if we wish to produce knowledge of everyday interactions and experiences in the contexts of mobility (see also Dossa 2008). Therefore, for us, resistance marks something that is not necessarily intentional and goal oriented, but something that comes into being relationally. Furthermore, the spaces of resistance are not necessarily conspicuous spaces designed or purposefully reordered to make the resistance omnipresent and noteworthy, as is often the case in research based on the legacy of Henri Lefebvre's writings on capitalist production of space (e.g., Harvey 2001; Soja 2010). For us, spaces of resistance can eventualize even in passing encounters, where the unwelcome body refuses to settle into the expected power geometry (Dikec 2005, 172).

For an understanding of the choreographies of resistance that are enacted by unwelcome bodies through their very appearance and visibility, the notion of abject is also of importance. A central concept for post-structural feminist, queer and crip theorization of the material politics of embodiment (e.g., Butler 1993; Shildrick 2002; 2012; Kafer 2013, 33), the term 'abject' literally denotes the state of being cast off, or utterly rejected as a vile disgrace of existence. Politically, the abjection of certain unwanted bodies from the prevailing orders is necessary, for the abjected play a constitutive role inside-outside the order. In the material and discursive practices of abjection, embodied others who deviate from the norm become included in the order only through their own exclusion. As constitutive outsides, those abjected thereby constitute the normative embodiments on which the order and its reproduction rely. In the neoliberal order of compulsory heteronormativity and able-bodiedness, for instance, those identified as sexually deviant or crip are multiply abjected from the zones of liveable life (e.g., Butler 1993; Shildrick 2002; McRuer 2006). It is then only through reiterated practices of abjection that the order remains normatively heterosexual and able-bodied, which makes the bodies of the abjected particularly threatening to the order, with their mere visibility presenting a potential rupture and a site of resistance.

Similarly, to the extent that the national order of things relies on the illusion of subjects as citizens incarcerated and territorialized to their particular place in the world, differentially mobile bodies through their very mobility threaten to disrupt the order, which is why these bodies are to be specifically surveilled. In this context, Imogen Tyler's (2013, 37) social theory of abjection as 'a mechanism of governance through aversion', as detailed in her book *Revolting Subjects,* is of great importance. Through a critical reading of Kristeva's work, and building on Judith Butler and a range of postcolonial theorists, Tyler argues for a historically contingent and politicized account of abjection that seeks to foreground the political potential of the abject to rupture the prevailing order. Studying the embodied positions and resistances of detained migrants, the 'Chav' (the white 'under-class' in the UK), young unemployed men and travellers in neoliberal Britain, Tyler shows how abject subjectivities are formed at the limits of the state, both within and at its territorial boundaries. She also demonstrates how this abjection of different types of 'scum' from the normative social order and citizenship helps to justify disciplinary measures that rely on the politics of austerity, eventually serving the global processes of capital accumulation.

For Tyler (2013, 38) then, governance through abjection is not limited to the mobile bodies of migrants only, and is not just a matter of border governance, but also relates to other forms of 'social stigmatization and social cleansing' that serve the neoliberal project. In this book too, the politics of abjection are understood as a politics of mobility, but the potential resistances of the abjected are to be found in the differentiation of mobile bodies, and in the creation of the categories of the wanted and the unwanted in choreographies of mobility. Citing Stefan Kipfer, Tyler also writes:

> Politicization of abjection involves 'territorial reappropriation' (Kipfer 2011, 1157). If the abject is spatializing politics of disgust, which functions to create forms of distance between the body politic proper and those excluded from the body of the state [. . .], then the politics of abject is a counter-spatial politics which attempts to reclaim the spaces and zones of abjection as radical sites of revolt and transformation. (Tyler 2013, 41–42)

For us, the counter-spatial politics of abjection takes place through bodies and in their corporeal choreographies of inter-relatedness.

Whether discussed in terms of abjection or something else, we maintain that resistance must be conceived as a bodily practice or a material event, which dislocates the focus on a speaking, sovereign subject or a group of subjects, such as social movements, as the initiator of resistance. The body is not, for us, simply a passive surface on which scripts of governance are inscribed, nor is it a surface 'that demands to be interpreted or written upon, by something other than the self' (Kirby 1997, 114). Rather, in our conception of choreographies of resistance, the governmental discourses within which subjectivities emerge are always already 'opened to an outside that reads, and outside that writes' (Kirby 1997, 127). This 'outside' of discursive power is the mobile body as it relates with embodied others, and power as it entangles with resistance is thereby always material-discursive (cf. Barad 2007). Our notion of resistance is therefore also related to the ways in which critique against the practices of governance is presented and how the body communicates itself. We argue that instead of one particular form or avenue of resistance, the multiplicity of choreographies that are used to communicate resistances should be taken into account. In this way, it is possible to grasp the rich variety of practices and events through which political existence is created, negotiated and articulated. Sometimes this necessitates moving away from understanding resistance as something that is related to wider social mobilizations and movements that are seen to politicize these resistances and taking minute gestures and interactions into account. Our exploration of choreographies of resistance involves engaging with such practices and events that cannot outright be recognized as resistance. For us, resistance has hence a strong connection to body, agency and politics.

In order to see the connections between resistance, body, agency and politics, we develop in this book *a practice-oriented and evented understanding of resistance* as a means to overcome dualistic notions of being and living in the world as well as to reject the 'romanticism of resistance' criticized by Sparke (2008) and the cultivation of hegemony noticed by Rose (2002). We question the ways in which people are categorized in hierarchical positions—through legal, political and economic material-discursive practices—that delineate particular spaces of agency for them. Such processes and practices of categorization tend to subsume agency. For example, Kimberly Dugan and Jo Reger (2006, 470) have suggested that in situations of seemingly 'weak' agency, people express feelings of hopelessness or uncertainty about making social change. This perspective assumes that people are aware of the goals towards which they want to and can aim. Sometimes, as in (prolonged) situations of uncertainty that are typical of some forms of mobility and that entail high emotional pressure, these goals might not, however, be articulate and clear. Therefore, it is necessary, in our view, to widen the horizon if we want to think about the complex relation between acting, being acted upon and togetherness in action.

Widening the horizon requires us to move from the verbal, textual and discursive to the corporeal and bodily (also Coole and Frost 2010). The ontological take on the body, which will be elaborated through the case studies, and the way that the mobile body matters enable us to examine those practices of existence that shape people's relationships to others and the world (cf. Coole and Frost 2010, 5). Taking into account the importance of bodies' materiality and relationality, political structures and the practices of governance targeted at bodies, means simultaneously challenging static notions of agency, and understanding that bodily motion does not occur in a fixed spatial and temporal framework, but it creates its frameworks (cf. Coole and Frost 2010, 19; also Manning 2007, 25). In short, we argue that resistance is a form of political enactment by which the body intertwines with the practices of governance in novel and innovative ways. This type of resistance is not always easily recognizable, visible or articulate. Resistance, as Steve Pile (1997, 3, cited in Takhar 2007, 133) claims, operates subtly in spaces not defined by those in authority and therewith more difficult to control.

GLOBAL (POSTCOLONIAL) POLITICAL ECONOMY AND THE GOVERNANCE OF MOBILITY

In this book, we are interested in events of resistance that create disturbance and find unexpected channels and ways for the expression of resistance. It is important to note that resistance always takes place and is connected with practices of governance as they are mutually emergent. The practices are not simply limiting, but they produce a space of oscillation between inclusion and exclusion as mobile bodies attend to power and practices of governance in multiple ways. Thus, it is imperative that we situate our notion of resistance within those practices and networks of power that are put into operation in the field of mobility. We position our approach through a Rancièrean take on politics and the political. The scale of practices of governance that is operational in the context of mobility varies from the global to the minutiae local instances where the human body is controlled and moulded.

As indicated in the introductory chapter, especially in the field of mobility, states are the key arbiters defining who is capable of becoming visible and belong to the community—which body is of value to the community and which is not. We thus concur with Veena Das (2008, 285) who notes that the 'nation-state sets up defenses to stave off the uncertainty emanating from dangerous aliens'. While doing this they create categories of desired and undesired bodies, subjected to a range of techniques of surveillance. Particularly when these technologies are discussed in relation to humanitarian migration, the mobile body emerges as a target of control, as the securing of populations against aliens is understood as requiring surveillance and accumulation and analysis of data concerning bodies and their behaviour (see, e.g., Andersson 2014; Bell 2006; Bigo 2002; Ceyhan and Tsoukala 2002; Dauphinee and Masters 2006; Dillon 2007; Dillon and Lobo-Guerrero 2008; Dillon and Reid 2006; Edkins, Pin-Fat and Shapiro 2004; Elbe 2008;

Guillaume and Huysmans 2013; Huysmans 2006; Huysmans 2014; Rajaram and Grundy-Warr 2007; Salter 2004, 2006, 2007; *Security Dialogue* 2008; Soguk 1999). These technologies and the spatial restrictions they enact are not, however, the same for everyone. In the racialized choreographies of global mobility, some bodies are more under surveillance than others, and this is also economically productive.

Modern nation states are repositories of ideas, images, ideologies where non-material and material practices and networks of governance intertwine (also Mountz 2010). The practices are not determined, that is, they are constantly created and struggled over. In the historical perspective, as Michel Foucault (1990, 138–143) has stated, particularly the techniques of control are constitutive of the nation state: the subjugation of bodies and the control of whole populations mark the production of nation states. Capitalism and the growth of nation states would not have been possible without the subjugation of bodies to capitalist production, control and the creation of population in order to serve economic processes (see Foucault 1977, 25). Simultaneously however, and perhaps paradoxically, global capitalism also relies on the differential mobility of racialized bodies, and hence on the rupturing of the nation state as a fixed entity. In their book *Border as Method, or, the Multiplication of Labor,* Sandro Mezzadra and Brett Neilson point out that, whereas goods and capital continue cross borders more fluidly than people, the inseparability of labour from living human beings requires that bodies remain in motion—when and in the ways the transnational labour markets demand. Mezzadra and Neilson (2013, 131–166) emphasize that the governance of mobility creates spatial zones of 'illegality' and 'legality', as well as temporal statuses of temporariness and permanence, that feed into the availability of cheap labour and, consequently, to the transnational accumulation of capital across and beyond nation states. This governmental play with the statuses, control, incarceration and emplacement of mobile bodies applies to migrants of all 'kinds'. Importantly then Mezzadra and Neilson's (2013, 137) account underlines that, although both research and policies tend to treat 'skilled', 'wanted' and 'legal' migration separately from 'unwanted, 'unskilled' and 'illegal', the mobile bodies in these different categories do not in fact 'occupy different universes' or live in some 'parallel worlds' of existence. Instead, it is exactly the differential 'controlling, filtering, and blocking' of mobile bodies that (re)produces the kind of flexibility of labour force that the expansion of global capitalism requires (Mezzadra and Neilson 2013, 21). They write about these entangled corporeal choreographies of controlling, filtering and blocking:

> States pursue the dream of correlating migratory movements with perceived economic and social needs established by statistical analyses of labor market dynamics, demographic studies, and political priorities. Systems of detention

> and deportation are crucial devices [...] [and] diffusion of skilled migration policies [...] seek to attune flows of migrants to the real and imagined needs of countries of destination. (Mezzadra and Neilson 2013, 138)

Indeed, as described in great empirical detail by Thomas Gammeltoft-Hansen, Ninna Nyberg Sørensen and their colleagues, entire 'migration industries' have emerged around these machineries of mobility, both for pushing bodies across the borders as well as for their detention and deportation (see Gammeltoft-Hansen and Nyberg Sørensen 2013; Andersson 2014). It is thus important to note that the choreographies of resistance examined in this book, albeit emerging in the apparatuses of different 'types' of mobility, are all entangled with one another, through a wider, discontinuous yet shared and ever-emergent context of global capitalism.

The order of global capitalism is racialized, and one important dynamics of motion in general, and for the entanglement of the choreographies of resistance in particular, is the tendency of bodies to tend towards 'whiteness'. This is a tendency that characterizes the global space of postcolonial political economy where the 'script of colour' is written on the human body. Whiteness is the norm against which everything (and everybody) else is to be compared, and for the normativity of whom the abjected make space through their own abjection. In the racialized order of global capitalism, the embodied positions and motions of the white European tend to dissolve into plainness (Keskinen, Tuori, Irni and Mulinari 2009; Loftsdóttir and Jensen 2012; Leinonen and Toivanen 2014). The white body is that which is 'at home' and need not move, or has the right to move with a minor amount of control and surveillance. As Sara Ahmed (2006, 136) puts it, whiteness is therefore that towards which everybody else coheres. With a reference to Franz Fanon's work, Ahmed elaborates:

> If the world is made white, then the body at home is one that can inhabit whiteness. [...] Bodies are shaped by histories of colonialism, which makes the world 'white' as a world that is inherited or already given. [...] Colonialism makes the world 'white', which is a world of course 'ready' for certain kinds of bodies, as world that puts certain objects within their reach. Bodies remember such histories, even when we forget them. Such histories, we might say, surface on the body, or even shape how bodies surface. (Ahmed 2006, 111)

In this book, we regard this postcolonial 'tending towards whiteness' as a globally relevant and shared dynamics of motion, of both resistance and governmentality, which ties together bodies and their differentially enacted choreographies across space and time. Although we will not engage with this coherence directly in all the empirical case studies, it remains a global, postcolonial undercurrent of (im)mobility, which we deem as important to recognize, even if and in particular when writing from the position of privileged whiteness ourselves.

ENTANGLED EVENTNESS OF RESISTANCE

All practices of governance as well as resistance produce material surroundings in which particular kinds of relations are created between differentially mobile bodies. However, particularly if governance and resistance are understood as mutually emergent dynamics that move with corporeal choreographies, the material surroundings are also always already shaped by resistance, and by bodies presumably subjugated by power (see, e.g., Rajaram and Grundy-Warr 2004; Epstein 2007; Norris 2000; Pugliese 2002). Despite their ambivalent space 'in-between' and of some bodies falling 'outside', mobile bodies, by their very nature, also challenge the meanings that are given to territorial borders and national populations (cf. Appadurai 1997, 4). Hence, they constantly challenge and (re)shape both national and global political order and the hierarchies that coincide with those orders. Thus, seen in the light of choreography, *the mobile body is itself a factor of turbulence and resistance*. By not quite fitting into the routines, flows and rhythms of the prevailing political and economical order, and yet not being fully separable from it, the mobile body produces dissonance and disruption within the order, thereby keeping the order itself in motion.

We argue that it is possible to discuss the actions, practices and the evented nature of the body's presence in terms of resistance, once the ontological context of the material body is taken into account (see also Coole and Frost 2010). Our framing allows us to think about the political and collective ethos of also 'private' and 'individual' expressions of resistance that would in many cases be deemed as uninteresting or irrelevant politically. The focus also allows taking note of people's lived realities and avoids presenting them as either victims without a voice, sheer targets of practices of governance, or incapable of articulating themselves. Through the different and internally variant case studies (chapters 3–5), we explore a range of intimate, presumably apolitical events and practices through which resistance is articulated, creating an understanding of multiple entangled choreographies of resistance. We engage with choreographies of resistance by paying attention to corporeality and those material practices that are created and transformed by practices and events of resistance. Rather than focusing on the actor, we emphasise action, that what happens (see also Orlie 2010, 117). Through a specific focus on *events* we address the social and political origins of people's conditions without necessarily having to rely on textual or verbal methods (see Dossa 2008, 13).

We use the notion of 'event' because events can be seen to have relatively vague spatial boundaries. In order to examine events in the context of mobile bodies and how resistance happens, we rely partly on Alain Badiou's (2005) notion of an event. For Badiou, events challenge the very constitution of the body politic: in events the foundations of the body politic are changed fundamentally. Events for him are gateways to future possibilities and reconfigurings, and they enable novel modes of being-in-the-world. Badiou is sceptical towards understanding anything in the domain of the state as resistant because the state tends to keep its grip on the situation. Politics within the frame of the state is, according to Badiou, only a matter of reactive protests which do not produce reconfigurings of the social reality. Although we concur with Badiou of the importance of events, we do, however, differ from his thought in a very fundamental manner: for us events are everyday occurrences, not rare revolutionary moments in history. In this, we return to an idea of community where events are disruptive surprises that 'dislocate' the community and the order of things (Marchart 2007, 69–75). Events rupture normality, the smooth functioning of disciplinary practices and, at the same time, open a space to rethink the hierarchical arrangements of the modern political and economical order.

In Badiou's thinking there is a connection between event and resistance, which is relevant for us too, as for him events are necessarily ruptural in relation to the dominant order. In our event-based approach, the focus is on spaces for politics, since a simple and clear-cut centre-periphery contradiction makes little sense when examining bodies on the move and their capacity for resistance (see also Mezzadra and Neilson 2013, 21). Approaching space and resistance with a performative and event-based approach, Davidson and Iveson (2014), for instance, look at the work of Jacques Rancière to think about the possibilities and practices of resistant politics. Davidson and Iveson's reading of Rancièrean politics highlights the importance of seeing politics as a spatial process, something that requires staging, mediating and subjectifying. Unlike for Badiou, for Rancière events are not necessarily revolutionary but also minute everyday occurrences, which play a significant role in introducing new political subjectivities and dissensus into existing social orders. They disrupt or disturb the prevailing order to which Rancière refers with the notion of the police.

Asking the question of how a singular and seemingly unformulated act of resistance might become something that initiates political change, Davidson and Iveson (2014, 141) argue that there are three spatial aspects to Rancière's thinking to which one should pay attention. First, there is the occupation of public space where neither space not occupation are in themselves political. This is because for Rancière a pre-designated political space does not exist. Rather, political activity is 'whatever shifts a body from the place assigned to it or changes a place's destination' (Rancière 1999, 30). Second, the politics of an occupation is therefore a question of subjectification and reinscription 'to the extent that they involve the subjectification of a "part of those with no part", a group that challenges the part supposed to be played by the bodies in that particular place in the process of declaring their equality' (Davidson and Iveson 2014, 142). Subjectification concerns the creation of identities that do not follow the existing hierarchies and roles. Furthermore, contestation becomes political through subjectification, that is, it becomes more than personal, and can transform existing orders (Davidson and Iveson 2014). Third, by mediations Rancière emphasizes, according to Davidson and Iveson, the imaginary dimension of space as the processes of diffusion and translation are critical to the fabrication of politics. A singular act of resistance can be translated and linked into universal demand for change through, for example, social media, and this can take place without pre-designated political media. In the circulation—made possible by techno-materialist developments in media—of moving images, videos shared on YouTube, place names and locations shared on Twitter interrelate and connect distinct acts and distant places (Davidson and Iveson 2014; for a techno-materialist perspective on media, culture and politics see Parikka 2012).

Unlike some scholars in the field of critical geopolitics argue (see, e.g., Swyngedouw, 2009; Holston, 2009), politics does not only emerge from the margin as if a structurally determined and excluded 'part with no part' became conscious and enunciated a demand that changes the society around it. Rather, the constituent elements of politics are the stakes of battle themselves. If resistance and governance are understood as being mutually constituted by bodies that come together in corporeal choreographies, as they are understood in this book, the dichotomic conceptions of agentiveness, such as 'margins' and 'the centre', 'the powerful' and 'the powerless', 'haves and have-nots', those with 'a part' and those 'without' become inconceivable. This is because in each event of corporeal encounter, trajectories of differentially mobile bodies bring together complex, spatially-temporally-materially diverse trajectories of resistance and governance, thereby enacting the particular configurations of governance and resistance that prevail at any one moment. In this conception, both resistance and governance are in constant movement—constantly in a stage of becoming—and this is because bodies are, by definition, mobile, material, and relational.

It is important to notice that for Rancière, politics is neither about reaching consensus nor about negotiating interests. Nor is it about distribution or procedures of consent. Rather, politics is about the de-classification of spatial orders that produces radical disturbance. As such, politics is a rupture within the present as those who are not visible with the dominant political order become visible. Political order consists of legitimized hierarchical practices that allocate particular (spatial) roles and possibilities (e.g., the citizen, the migrant, the undocumented and the skilled worker) to various members of that system. The members might have different roles in different practices, yet the roles are hierarchical. These hierarchical arrangements are something what Rancière calls *police*. At the core of the police there is the inequality of the constituting members. *Politics*, which differs from police, is a matter of knowing who is qualified to become visible as 'politics revolves around the properties of spaces and the possibilities of time' (Rancière 2003, 201; also Rancière 2001, thesis 7; Rancière 1998, 28–29; Rancière 2006, 13). For Rancière, politics is thus about disturbing the arrangement of roles and hierarchies within societies. He writes:

> Political litigiousness/struggle is that which brings politics into being by separating it from the police that is, in turn, always attempting its disappearance either by crudely denying it, or by subsuming that logic to its own. Politics is first and foremost an intervention upon the visible and the sayable. (Rancière 2001, thesis 7)

In politics, reconfigurings of the existing orders declassify and undo the supposed naturalness of orders and their hierarchies. Politics replaces the existing orders with controversies and divisions. From such a position, when political agency emerges from those who have not been visible, the questions 'who are we?' and 'where are we?' arise (Rancière 1998, 99–100; 2009, 116). When the constitutive outside that is not accounted for in the order of the nation becomes visible, the outside opens up the possibility, if not the necessity, for the enactment of the political as it composes itself through contingent, inventive and interruptive combinations. *Politics in this view is hence a particular type of relational event that emerges with respect to and is intertwined with the police orders.*

In sum, the Rancièrean view of politics becomes especially relevant for our take on resistance that we outlined earlier and that not only implies a protest against something, but evokes a struggle at the foundations of political existence and expression. The empirical cases that we address in this book, illustrate that the mobile body can disturb the practices of governance and it bears agentive capacity and enacts political agency despite of it being categorized unwanted or wanted, being in between or even outside. We concur with Rosalyn Diprose et al. (2008, 274–275) who argue that an exclusive focus on practices of governance fails to articulate the ways both human and nonhuman material life contest those regimes of governance and control. In a similar vein, we argue, and illustrate in our case studies, that the networked practices of governance, the political order (*police* in the Rancièrean sense) that those practices create open up spaces for the situational choreographies of resistance. They create 'body-spaces' that are always ambiguous, *inter-zones*, spaces that are not only constraining or conditioning, but also enabling and transformable (cf. Puumala and Pehkonen 2010). We hence illustrate in this book how variously categorized mobile bodies articulate their (ontological) agentive capacity and how their choreographies of resistance bear potential for disrupting the prevailing order.

THE POLITICAL POTENTIAL OF MOBILE BODIES

The emphasis that we place on the body and corporeality dislocates agency as the property of the sovereign subject (also Coole and Frost 2010, 20). For us, the mobility of the body represents a corporeal attempt to make sense of and negotiate physical space through bodily action. In this respect, we again have a lot in common with Diane Coole and Samantha Frost's (2010) ontological take on the body when they write:

> Bodies communicate with other bodies through their gestures and conduct to arouse visceral responses and prompt forms of judgment that do not necessarily pass through conscious awareness. They are significant players in games of power whenever face-to-face encounters are involved. (Coole and Frost 2010, 20)

Movement shapes and changes the body and thus the body represents the opening of space or the events of space being formed (also Puumala and Pehkonen 2010). The spatiality to which we refer is both the 'spatiality of position' and the 'spatiality of situation' (Connolly 2010, 182). These different realms of spatiality affect the way in which the body matters both to the person himself or herself and how others perceive that particular body. They also affect the ways in which the body can move within space. The take on the connection between corporeality and space bears some important consequences with regard to how resistance, agency and the political are understood in this book. The adopted understanding highlights being as it happens, which again undermines the idea of substantial identities, centres and peripheries and stable hierarchies between variously located and categorized bodies. Like choreography, the constant unfolding of being can be partly pre-written in terms of the movements that the person can enact, but at the same time it is only lived out and takes shape in the movement that in the end constitutes the choreography itself. Through the mobile body, it is possible to explore alternative ways of knowing, thinking and being human, and to picture agencies and solidarities beyond the constraints of various seemingly obvious networks of power.

The body is vital as it disposes us outside ourselves and sets us beside ourselves. The body, in Judith Butler's (2004, 26) words, opens up another kind of normative aspiration within the field of politics. It possesses an invariably public dimension (Butler 2004, 26) and that is what makes it political. In this understanding, the body itself is a space that through its gestures, connectedness to other bodies and subtle external and internal movements reaches and articulates the space of the political (also Manning 2007, 58). This type of corporeal politics lies in contradiction with a linear, territorialized and bounded conception of space proposed by, for instance, the principle of sovereignty (Manning 2007, 59). Even though it cannot be claimed that practices of categorization would determine the possibilities and capacities of the body, it yet is the case that those practices orient bodies in some—and not in other—ways (also Ahmed 2010a, 247). Spaces are oriented around bodies, claims Sara Ahmed (2010a, 235) as they show that some bodies are more 'in place' in some places than other bodies (also Cresswell 1996). In such a context, corporeal choreographies of resistance can be understood to point towards the future: the way in which bodies are oriented affects their tendencies and their resistances point to 'what is not yet present' (Ahmed 2010a, 247).

Much like our take on resistance, the notion of the body that we propose, is characterized by openness and eventness. Such a take suggests that bodies resist exhaustive telling and knowing. It conceives of the body as a sense in action, not as something that makes sense with regard to a particular political order or category. Furthermore, it is important to acknowledge that people on the move often refuse to accept the categorization of their bodies created through the practices of governance, and protest against it through gestures, gait and miniscule moves. Thus, relations of power always not only entail resistance, but also leave space for contesting the working of that power and acting against it. Taking a cue from the idea of the material body as leaking, expanding and reaching out, political agency can be understood to unfold as not only multilayered but also infused with contradictions, ambiguities and ambivalences. What we aim to imply is that instead of a single way of acting politically, there are shifting strategies of agency and interactions of resistance that are characterized by constant renegotiation of the field of action and of one's positions.

Choreographies of resistance are closely connected to the dynamics of materialization. By this we intend that it is imperative to address the dynamics through which subjectivities are created and examine critically the domains of power so as to rethink the dimensions of agency (see Coole and Frost 2010,

37). Bodies do not always have ample space to politically voice their experiences and views—their ways of enacting agency are fluid, shifting and conditioned by governmental policies. Becoming located in the categories (citizen, labour migrant, undocumented migrant, tourist, etc.) available, bodies on the move are situated within the political, social and economical order in a way their potential agency becomes in its implications less threatening for the places that they inhabit (cf. Zetter 1991, 57). Yet, bodies on the move remain within the sphere of the political because they are involved in processes of subject formation and forms of agency that are beyond the grip of some statist governmental policies (cf. Moulin and Nyers 2007, 362). In other words, the space of agency is not always readily available for people, but by engaging with and relating or refusing to relate to other bodies, they create the potential for politics. The potential remains always open and subject to change. An emphasis on materialization and eventness leads us to focus on the body's inherent capacity to relate to other bodies. In that understanding, the body has a central role in politics, but further yet, it implies that the body is possibly inseparable from the political.

The political capacity of the body emerges from its materiality and movement that both, to a large extent, are beyond the control of networks and practices of power (cf. Grosz 2010, 141). The mobile body, at some moments, may be capable of addressing and relating with its material surroundings and situationality in a way that represents its capacity of being otherwise. We take after Elizabeth Grosz (2010, 151), who claims that 'the question of freedom for women, or any oppressed social group, is never simply a question of expanding the range of available options so much as it is about transforming the quality and activity of the subjects who choose and make themselves through how and what they do'. We suggest that through mobile bodies and the relationality of our existence their resistances make us encounter, we can begin to understand how people constitute themselves through how and what they do. And, how what we are is never a matter of one body alone, but is about the ways in which bodies orient towards and are vulnerable to one another (see also Braidotti 2010; Despret 2013, 61, 69; Degnen 2013).

Chapter 3

Postcolonial Space and Entangled Corporeal Choreographies

In this chapter we discuss the corporeal choreographies of resistance that emerge in the global labour markets of care.[1] Taking place in the discontinuous yet shared space of the global political economy, we argue that the recurrent racialized categorization of mobile bodies and their value is to be spatially understood as entangled corporeal choreographies. In these entanglements, some forms of non-conformation may reiterate rather than challenge the racialized orders on which global neoliberal capitalism relies. Therefore, we emphasize that analysing the corporeal choreographies of mobility in the context of juridically or otherwise different 'groups' of mobile people is not enough. It is equally important to consider, how the different corporeal choreographies of resistance are entangled and move with(in) each other, sometimes in contradictory ways.

With a particular focus on Filipino nurses as a global care reserve and exportable mobile labour, we begin with an analysis of the Philippine state discourse of labour emigration as a technology of governing mobile bodies from a distance (e.g., Guevarra 2010; Rodriguez 2010). The corporeal choreographies enacted in this context are deeply interwoven with the postcolonial trajectories of global capitalism that necessitates the racialized division of bodies into categories of differential value (cf. Mezzadra and Neilson 2013). Empirically, the chapter begins with the context of the Philippines, focusing on the ways that the bodies of overseas Filipino workers (OFWs) in general and nurses in particular are governed from a distance. The global mobility of the caring Filipino labour is facilitated by the world's most highly advanced state machinery, designed for the effective governance of *out*migration (e.g., Guevarra 2010, 90–91). The governance of OFWs as globally mobile bodies cuts through the entire state machinery. Yet, it does so in a completely different manner than in the European understanding of 'governance of migration', where the state power is understood in terms of border controls, capacity for efficient surveillance practices (including profiling, screening, etc.) and, ultimately, keeping the unwanted migrants out. In the Philippines, citizen bodies are governed so that they choose to migrate across the world in a managed fashion. The chapter shows how the seemingly disparate, and differentially valued, trajectories of mobile bodies come together unexpectedly in complex webs of governance and resistance, where whiteness remains as that which coheres, and tending towards whiteness as that which mobilizes (cf. Ahmed 2006).

Towards the end of the chapter, we turn to an example of a concrete, embodied and implicitly racialized event of elderly care, suggesting that perhaps the body as a living, dying, relational and dependent organism always already resists the racialized choreographies by which the political economies of migration are presently being orchestrated. The chapter demonstrates that corporeal relationality is always already there: in our inescapable need of connection with other bodies. Yet, although the body resists, such choreographies of corporeal connectedness do not completely dismantle the historically inscribed ethnic and racialized hierarchies that concretely materialize in our (postcolonial) lives. The body brings them forth and makes them visible. In short, we argue that it is important to recognize how the material body ultimately refuses to comply with the historically inscribed divisions of wanted and unwanted bodies, white and non-white.

THE PHILIPPINES AS THE GLOBAL LABOUR RESERVE

The Philippines is known as one of the world's central labour reserves, where the state actively promotes emigration (Guevarra 2010; Rodriguez 2010). As a systematic state practice, policies of transnational labour brokerage were first introduced by the Marcos regime in the 1970s, as a response to the growing unemployment and debt crisis (Guevarra 2010, 31 ff.). Although at the time labour exportation was meant to be a temporary solution to reduce budgetary deficits, it has over decades become the *sine qua non* of the national economy. With high external government debt, and the country capable of paying back little more than the interest each year, the Philippines has grown highly dependent on the remittances that the overseas migrant workers send back to their families (cf. Bello, Guzman, Malig and Dogena 2005, 14–15). In this case, global mobility is actively promoted by the state, for economic reasons: for the national economy to grow, the embodied labour of the citizens must continue to move out in increasing numbers.

Filipinos migrate for work in all types of professions. Women leave as domestic workers, nurses, and care workers, for instance, men as plumbers, engineers, electricians and construction workers, and both as cleaners, waiters and cooks, and also in professional categories which are labelled as 'high skilled'. Every third seafarer in the world's shipping industry is estimated to be a Filipino (Department of Foreign Affairs 2014), meaning that much of the global logistics is enabled by the embodied labour of a Filipino (cf. Sampson 2003; McKay 2007; also Chua 2015). In research as well as in the popular imaginaries, however, the global labour force of Filipinos is particularly known for the feminized labour of domestic workers, caregivers and nurses (e.g., Buchan 2006; Lopez 2012; Madianou 2012; Onuki 2011; Parreñas 2001, 2003, 2005, 2008; Näre and Nordberg 2015; Tungohan 2013; Vaittinen 2014; Vaittinen et al. forthcoming). This feminized workforce represents the 'national trait' of 'Tender Loving Care' of the Filipino, that is, 'TLC' in Philippine governmental jargon and popular parlance, and not without a reason. Different types of care work remain the biggest professional categories in which Filipinos migrate. In 2014, for instance, out of the total of almost half a million new contracts recorded in all land-based professions,[2] over 180,000 contracts were for household service workers, over 19,000 for nurses and 12,000 for care workers (Philippine Overseas Employment Authority 2014), most of whom were women. As the ageing of the populations in the post-industrialized 'West' increases the global labour deficits in elderly care—that is, in a niche in the labour markets that is racially categorized as 'migrants' work'—the deployment of Filipinos to overseas care work is also likely to increase in the future.

We argue that the global mobility of the Filipino population must be read in the context of the postcolonial political economy as indicated in chapter 2, and its counterpart, the (inter)national order of things. The debt burden of the state is directly related to the neocolonial economic policies imposed on the country by international financial institutions, which emphasize export-oriented economic structures at the cost of self-sustainability, and structural adjustments at the cost of social security (cf. Bello, Kinley and Elinson 1982; Bello et al. 2005). Furthermore, these mobilities have historical roots as the national tradition of labour emigration dates back to the US colonial practices (Rodriguez 2010, 1–10; also Guevarra 2010). For instance, the utilization of Filipino men as plantation workers in Hawaii in the early twentieth century, the US-Philippine military collaboration, and a series of guest worker and educational programmes have created a national history and 'culture of migration' (Choy 2003, 4), while gearing the Filipino 'hearts and minds' to dream about a life overseas (Rodriguez 2010, 5). Similarly, the institutionalization of the American educational system and nursing curriculum in the Philippines has contributed to the construction of the Philippines into a global nurse reserve (cf. Choy 2003; Guevarra 2010: 29–30; Ortiga 2014). It has been estimated that, over the years, up to 85 percent of nurses educated in the Philippines at all times have migrated for work (Lorenzo et al. 2007, table 1). This is largely the result of the active nurse brokering policies of the state. Since the early 2000s, the Philippine government along with its largely privatized machinery of higher education has heavily invested in its capacity to provide nurses for the increasing 'global demand' (Ortiga 2014; Masselink and Lee 2010). At the moment, as a combined result of governmental miscalculation, inadequate governance of private universities, and the global financial crisis in 2008, there is now a docile population of over 200,000 university-educated registered nurses in the Philippines, willing and able to migrate for work anywhere in the world (cf. Vaittinen et al. forthcoming).

RACIALIZED HIERARCHIES OF BODIES: 'NOT QUITE THE SAME AS THE REST'

Filipino Diplomat A: It's really a very interesting study, in terms of a—migration of workers, you know. Unfortunately, when you say—*migrant worker—there is a negative allusion to it.* Right? *It's like you're on the lower level all the time.*

Tiina: I, I … but that is also, like, one of my ambitions that I'd like to, see the other way around.

[…]

Filipino Diplomat A: Because you know, in the media, they focus completely on the Tunisian people on the ships. Because this is what makes news! And they forget that—we have a Senior Manager for example in Nokia. Technically, she is classified [...] is a migrant worker. You know. Because she's out of her country, but…She's a Senior Manager in Nokia, for the last five years. Even among themselves—I don't think she sees herself as a migrant worker. What she sees herself is...

Tiina: professional…?

Filipino Diplomat A: … an expat.

Tiina: Exactly.

Filipino Diplomat A: Do you know that if, in most of the countries I'm going to, if I as ambassador needed an intelligence network, inside household(s), of the American, (-) and the EU countries, *I have it.* But we don't use it of course. We have people working, as far as … The nanny of Prince William is Filipino. […] *So that's how far in we are.* [...] And these are domestic, OK, *so, do you call them migrant workers in the same way as this migrant worker who's jumping off a ship?* It's completely different.

Filipino Diplomat B: In fact if we look at some of, in some of the countries, like in Brunei Darussalam, there are more Filipinos than the Bruneians themselves, because of their keeping of all the economy, because, our people, like managers, cleaners, they're the ones running the place.

Filipino Diplomat A: There's, some countries are *completely dependent on us* to run their economy. So, this is the. ... It's a different picture altogether. *We're not quite the same as the rest.*

(Interview, Tiina, embassy of the Philippines in Helsinki, May 2011, our emphasis)

The vignette above consists of two extracts of a taped interview, conducted with two diplomatic state representatives, at the embassy of the Philippines in Helsinki, May 2011. As part of Tiina's multi-sited ethnography of the global political economy of care and Filipino nurse migration in Finland (Vaittinen, forthcoming), the purpose of the interview was to learn about the impact of international nurse recruitment on Finnish-Filipino state relations. The discussion, however, ended up taking its own routes. The extracts result from a somewhat misinformed interview plan where the researcher, at the very beginning of her research project, accidentally used the wrong terminology when referring to Filipinos working abroad. Whereas in the Philippine state discourse the official term would have been OFW, in the interview plan sent to the diplomats beforehand, Tiina had recurrently used the term 'Filipino migrant worker'. This, and particularly the term 'migrant', the interviewees read as derogative, although it was not intended as such. This unintentional provocation then led to lengthy discussions as the interviewees educated the interviewer on the negative connotations of the term 'migrant', caused by the unwanted border-crossings taking place in the Mediterranean.

This research encounter can be read as opening up the eventness of the political where bodies and their choreographies connect in unexpected ways. First, opposing the interview plan's tendency to speak of OFWs in the same category as those 'jumping off a ship', the Philippine state representatives saw it as necessary to explicitly draw the researcher's attention to highly skilled Filipinos in Europe, such a female senior manager in Nokia. She, the interviewees emphasized, would barely 'see herself as a migrant worker' but rather, as 'an expat'. The diplomats' need to convince the researcher about the capacity of their nation to produce not only 'migrant workers' but also 'expats' reflects the predominant discourses, where overseas work is hierarchically ordered in terms of 'skills' as well as gendered, ethnicized and racialized postcolonial legacies (cf. Kofman and Raghuram 2006; Kynsilehto 2011a; Mezzadra and Neilson 2013). Thus, the work invested by the diplomats in inscribing the OFWs as something a bit better than any other workers and migrants, can be read as a form of state action against the postcolonial racialization of its people in the global discourses of mobility. Secondly, in this mundane event of an interview, the researcher was thrown to a site of researching mobile bodies that was completely different from the site where she imagined to be. Tiina entered the interview with an understanding that she was studying Filipino nurse migration as a particular, care-related form of labour migration to ageing Finland, as embedded in the EU. Yet, she left the interview with a puzzled realization that the choreographies involved in her study were clearly not only about the embodied mobility of Filipino nurses, or about 'labour migration'. Instead, in the postcolonial spaces of the global political economy in which the research took place, the very mobility of 'the Filipino nurse' was connected and interwoven with the (im)mobility of a variety of other bodies as well as with bodies' relationality and agentive capacity.

A long while after the interview, Tiina remembers thinking about Henning Mankell's Tea-Bag as well as Anitta's work in the Mediterranean and in the 'Jungle' of Calais (see chapter 4), pondering about the potential connections—theoretical, discursive and material intertwined—between two very different 'types' of mobile bodies, those of 'the labour migrants' and those of 'the paperless'. She also remembers a discomforting feeling of numb guilt right after the interview: the sense of having her own embodied position as the white Nordic/European observer oddly dissolved in these choreographies, where the mobile body to be observed tends to be that of someone else (cf. Leinonen and Toivanen 2014; Loftsdóttir and Jensen 2012; Keskinen et al. 2009). In her field diary, immediately after the interview, Tiina reflected on this:

> But where does it leave me … ? Not a migrant worker, not here, not now. Not a Great Filipino Worker, a body that's a bit better than the rest. Not a Great Worker, definitely not a great ethnographer, for I do not find my place here, my relation…Not this, not that, just Plain Worker. In this national order of things, this time and place, I'm obviously always better off than the Filipino, who needs to be defined as better than the rest because s/he cannot just be plain. I am the white body that is just plain, and as such always better off.

Tending towards whiteness and opposing the racialization of their population in the European labour markets, the Filipino diplomats Tiina interviewed found it necessary to emphasize, recurrently, that the Philippines is not just another nation of 'migrant workers', but something else: *a little bit better*, not quite the same as *the rest.* As elaborated in the previous chapter with a reference to Sara Ahmed's notion of whiteness as that which coheres, in the global, postcolonial choreographies of mobility, the embodied motions of the White European tend to dissolve into plainness in the sense of a norm. The white represents that against which everything else is to be compared; for the normativity of whom the abjected make space through their very abjection. Indeed, were Tiina interviewing the diplomatic staff of some 'Western' ('white') state about its population working abroad, she might have herself used the term 'expat' as opposed to 'migrant'. The 'expat' is, after all, the usual label attached to a 'Westerner' working abroad. However, particularly if the job is categorized as 'low-skilled', or feminized like care work or nursing, it is rarely discussed as a job for 'expats'. Then it is a 'migrants' job', both discursively as well as in the actual material patterns of employment (cf. Dahle and Seeberg 2013; Olakivi 2013; Näre 2013a; Wrede and Näre 2013). Indeed, in the context of the prevailing discourse of 'migrancy', which places racialized workers in Europe in a lower social category (Näre 2013b), the diplomats' scripts become understandable, if not acceptable.

Describing the aims of the government to perceive the OFWs as valuable expats, the diplomatic staff inscribed the bodies of Filipino migrants with a valorized difference. This was done by the diplomats because this is what the state machinery and its police order (in the Rancièrean sense) expects them to do, that is, to know who is qualified to become visible and administer the ways visibility is possible. A pivotal tool in this practice is the conceptualization of OFWs as 'modern day heroes', the *bagong bayani.* The *bagong bayani* is a figure of the Great Filipino Worker that is constitutive of the Philippine national order of things. The term was first coined in 1988 by President Corazon Aquino in her speech to Filipino domestic workers in Hong Kong (Guevarra 2010, 33; Rodriguez, 84–85), as a 'morale booster' to the overseas migrants (Diplomat A, interview in Helsinki, May 2011). According to Guevarra (2010, 333), naming the overseas Filipino workers as heroes reflects the government's desire to summon 'a kind of nationalist spirit rooted in the belief of one's role in nation building'. In the Philippines, the figure of the *bagong bayani* is constantly reproduced and reiterated in the public discourse: in the media, in the recruitment advertisements, in the compulsory pre-departure orientation seminars of OFWs and many other government policies. As such, the figure steers the ways overseas work is valorized in national imaginations. A closely related concept to *bagong bayani* is *balikbayan,* which translates from Tagalog as 'nation returnee' or 'one who returns to the nation'.

The figures of both the *bagong bayani* and *balikbayan* are present everywhere in the national discourse. As positive and desirable figures, *the 'heroic migrant' discourse becomes internalized in the OFWs' embodied lives*, in the ways that the globally mobile Filipino workers think and act in the world, and represent themselves to it (cf. Guevarra 2010; Rodriguez 2010). Overseas work often means long periods of separation from one's children and other family members, and is therefore rarely just the life of dreams. However, when the hardships to do with mobile life are understood as heroic—that is, when the normally derogated migrancy becomes coined as heroic—the choice to leave gains a higher value: it is a sacrifice for a better future, for oneself, for the family and, eventually, for the nation (also Guevarra 2010; for resistance, see Tungohan 2013). In Tiina's research for instance, the individual nurses participating in her study rarely represented their decision to leave as a sacrifice for the nation, but rather for the family and the self. This, however, makes the *bagong bayani* and the *balikbayan* discourses no less effective for the state, governmentally, for neoliberal governmentality tends to operate through self-care in individual lives and gendered family relations (cf. Foucault 2008).

ENTANGLED CHOREOGRAPHIES OF MOBILITY AND GOVERNANCE FROM DISTANCE

In the case of the Philippines, the question of border control and state governance of migration (and thereby the question of resistance) gains new meanings, when compared to the European context of managing *immigrant* bodies that are discussed in chapters 1 and 2 and later in chapters 4 and 5. Yet, emerging in the shared space of global (postcolonial) political economy, the choreographies of bodies on the move and migrant control are entangled in crucial ways. First, both the unwanted migrant body in Europe and the heroic OFW are politically evaluated in primarily economic terms. Compare for instance, the African 'migrant' and the migrating not-quite-white Roma beggar as a burden to the European economy against the OFW as economically profitable for the sending state, and in the long run for the receiving state too. Secondly, these economic evaluations, and the mobilities in which they are enacted, are mutually constructed, as non-white and not-quite-white bodies are made to strive towards the position of whiteness. None of this, however, arises naturally. The corporeal choreographies and their entanglements emerge as bodies are inscribed with differential values, positionalities and agentive capacities. These (re)inscriptions are not simply about writing bodies but also and in particular about writing bodies differentially, so that the relations between bodies remain hierarchical, and the motion towards whiteness re-occurs. Read in the apparatus of the postcolonial global political economy, corporeal choreographies are enacted through the writing of difference and power relations in and between bodies; through the (re)writing of certain bodies as potentially present and potentially mobile, while actively abjecting others.

For instance, as part of the interview structure prepared for the diplomats, Tiina had listed a range of terms that describe Filipinos as global workers in the literature, but that can in some contexts be perceived as politically controversial labels. In the interview, the diplomats were then asked to reflect upon these terms and their meanings from their own positions and perspectives.

One of the terms listed was *bagong bayani,* also referred to as the Great Filipino Worker. This term immediately returned the interview back to the question of 'migrants' in the Mediterranean:

> *Tiina:* How about the term 'the great Filipino worker'—I think it was the former president that launched it?
>
> *[...]*
>
> *Diplomat B:* Ehm. ... The idea of, or the concept of the hero, the modern day hero ... *calling them modern day heroes, because of the contribution they give to the country.*
>
> *Diplomat A:* I also think it was part of this—because we realized that the concept of migrant worker was taking a negative connotation, because of the other ... because of the *exposure* of the others. *You have Somalians, you have—you know. These people.* And ... we honestly feel that our Filipino worker overseas is—a little bit more qualified. And, recognition to—of that qualification and his contribution to the Philippines needed to be recognized. That was the basis, of calling them a hero.
>
> *Tiina:* Right [pause]. That opens—yeah that opens it up a bit.
>
> (Interview, Tiina, embassy of the Philippines in Helsinki, May 2011, our emphasis)

The literally non-diplomatic language, or the presumption of the diplomats that 'those Somalians' can be discussed in this manner with a researcher, exposes the ways in which bodies are inscribed with a difference. Drawing a difference between Filipinos and black African bodies through their exposure, this extract also implies how colonial pasts are inscribed on bodies, and how those histories 'surface on the body, or even shape how bodies surface' (Ahmed 2006, 111).

In the global postcolonial economy, and in its choreographies where bodies tend towards whiteness while relating with other bodies, some colonial pasts and relations are more valuable than others. At the level of representation, the postcolonial relationship between the Philippines and the United States, for instance, can be and is utilized profitably, but perhaps only in relation to 'other others' (cf. Ahmed 2006; 2010b). Being once constructed by the colonizer as an 'empire of care' (Choy 2003) is something that is explicitly utilized in the marketing of Filipino nurses to the world, not only by the state but also by the business actors involved in the labour brokering. For instance, in an anthology titled *Managing Mindsets: Applying the Cultural Advantage in your Business,* a cross-cultural consultant working for a transnational recruitment business in Finland describes positively the ways that 'Filipinos do not seem "that Asian"', due to 'three hundred years of Spanish colonisation, followed by almost fifty years of American occupation' (Rogers 2011, 85). Valorizing this positive cultural difference in the context of Finland, Rogers emphasizes that, to make the most out of this difference in business, the Filipino nurses are not meant to

> turn into Finns, which brings the cultural advantage of the Philippines to Finland. They are seen as remarkably friendly and warm. Obviously, this can also be said of many Finns, if only one knows what kindness and warm-heartedness look like in Finland. Finns themselves appreciate the friendly attitude of the Filipinos. The Filipino nurses are talkative, caring and warm towards patients and colleagues, and it comes 'naturally'. (Rogers 2011, 87–88)

Thus, the Tender Loving Care of the hard-working Filipino is not only an advantage for the remittance-dependent state, but also for the recruitment business and for the employers in Finland, who can provide better care by hiring an Asian who is really 'not that Asian', thanks to the (right kind of) colonialism. What we wish to emphasize here are the entangled corporeal choreographies at play. When the Filipino migrant worker is constructed as heroic, it is notable that this is not mainly done against the plain white 'expat' holding the power in the centre of the national order of things, but rather against certain abject bodies in the very margins of the order; against those with whom no one wants to be identified. Perhaps, the Philippine state governs its migrant bodies as heroic so as to smooth the entry of its population to the racialized transnational labour markets, which are often explicitly racialized as the citation of a business actor above shows. Yet, this valorization of the Filipino bodies is done in ways that the racialized order itself is not much, or is only minimally ruptured. The strategies that the state of the Philippines applies are meant to permeate the same racialized boundaries, which Tea-Bag and other 'boat people' push through, while risking their lives doing so. Yet, the Philippine strategies of governing from a distance pushes the likes of Tea-Bag further away from the shore, while making way for the bodies of the *bagong bayani*. It is as if the (im)mobility of Tea-Bag, and those whose dead bodies at the bottom of the sea are starting to form a bridge across the Mediterranean, were governed through the corporeal choreographies of the OFWs.

From the perspective of corporeal choreographies, in these subtle processes of governance, and not so subtle inscriptions, the bodies of the OFWs become mobile nodes of governance in contradictory ways. On the one hand, as heroic global labour that is better than 'the rest', the bodies of the *bagong bayani* are produced as better than the other brown or black bodies that migrate. In order to keep up the global motion, the Filipino body must continue to meet its heroic expectations. The state reaction to the postcolonial racialized scripts of its people as lesser than the 'white' requires that the mobile Filipino bodies remain docile vis-à-vis the state governance. On the other hand, however, these governmental scripts written on OFW bodies are effective only insofar as they continue to be rewritten, reiterated and reinscribed. To the extent that all migrant trajectories are always unique corporeal paths across borders and through global space, each OFW trajectory is also a potential rupture of and resistance to the state governance of emigration (cf. Vaittinen 2014).

RELATIONAL AND MATERIAL BODIES—RESISTANCES FROM WITHIN?

We concur with Mains et al. (2013, 2) who argue that 'the bodies of [...] postcolonial migrants continue to provide a daily reminder of the spaces and practices of colonial pasts and the necessity for a critical understanding of the postcolonial present (and future).' Yet, our research demonstrates that the corporeal choreographies of mobile bodies intersect in unexpected ways in particular space-time configurations. When we start to understand the material-relational body as resisting the hierarchical scripts that the (inter)-national order of things inscribes on its surface, we will soon notice that the migrant's body is not the only body on which the order is being inscribed, for the inscription of bodies, like all writing, is a relational process of differentiation (Butler 1993; Derrida 1980; Grosz 1997; Kirby 1997). Furthermore, although this book largely deals with the resistances enacted by migrant bodies, we would like to think that the migrant's body is not the only (im)-mobile body that resists the order. Since migrant bodies, as any other bodies, are relational bodies, namely bodies that move in space while making space political, they have capacity to enact political choreographies anew (Puumala and Pehkonen 2010; Puumala et al. 2011; Vaittinen 2014; Vaittinen 2015) when they extend and reach out towards others. This, we think, is an important aspect to examine, lest we might just be reproducing the dichotomy between the subaltern body of 'the migrant', and the speaking subject of the citizen (cf. Spivak 1988). Let us therefore turn, for a moment, to the concrete corporeal relations of care, and their potential meanings for the postcolonial, material-discursive hierarchies constructed in our minds and through our bodies over time.

In Finland, the international recruitment of Filipino nurses is still a relatively new and small-scale phenomenon, dating back to 2008, when the first Filipino-registered nurses were recruited by private companies, to work in mainly private care homes for the elderly. Since then, approximately three hundred Filipino-registered nurses have arrived in Finland to work, primarily through complex and lengthy recruitment programmes conducted by two separate companies and their partners in both Finland and the Philippines. The details of the programmes vary, but they usually begin with an intensive ten to fifteen months of Finnish language and cultural training in the Philippines. Consistent in these recruitment practices is a pattern of deskilling, indirectly institutionalized by Finnish law. Namely, in Finland, it is legally impossible for third-country nationals (those coming from outside the EU) to get their qualifications recognized in the healthcare professions. As a consequence, the heroic Filipino-registered nurses, with American standards of education, are not recognized as professionals by the Finnish/EU legislation. Instead, they are treated in Finnish immigration and labour market policies like any other 'migrants'. Having no chance to practice their profession as registered nurses without further (Finnish) education, the Filipino nurses in Finland are contracted to undertake menial care duties, as practical nurses, nurse assistants, and even cleaners (for details, see Vaittinen et al., forthcoming).

In this context, one of Tiina's ethnographic field sites was a vocational school, where a group of some thirty (mainly university educated) Filipino-registered nurses studied for two and a half years to become Finnish practical nurses.[3] This vocational degree programme was for the nurses an entry point to Europe and to the Finnish labour markets, with routes of direct recruitment having become impenetrable for one reason or another (cf. Vaittinen 2014). In the following section, we sojourn through that school, where the highly skilled OFW nurses train to become equal to Finnish nurses with a lower level of education.

> Today, I was visiting the campus of the vocational school where my research participants study to become practical nurses in Finland. Most of them are registered nurses in the Philippines, with four year university degrees, board exams passed, and usually also with clinical work experience in different types of hospital environments. Now, they are here in Finland, studying for two years to become qualified practical nurses in elderly care—a job where they will not be able to utilize all their clinical, professional skills. They always tell me, how the opportunity to migrate through the programme is really a fantastic opportunity—however, the practical nurse's degree programme itself is boring and brings them nothing new. Only the language (which they learn for only 2–3 hours per week) is difficult, but otherwise it's mostly about basic care tasks: stuff that they know, but cannot speak about in our language(s). To me, however, the classes are interesting.
>
> I got the opportunity to sit in the class again today, to observe and participate. The topic of the class was: 'how to stimulate the senses of elderly people by bodily means, talk or touch'. Super! I wrote down notes as I listened, even ended up commenting on something. Given that my research is about the corporeal relations of care in the global political economy, it was highly interesting to learn about caring as a concrete, material, bodily relation. Unlike in most of our

social scientific theories, in the class it was obvious that the corporeal relation of care is a physical as well as a sensual relation; that there is more to us as human beings than our rationality. In my notebook, I wrote down the teacher's words:

- By stimulating different senses you can give the elderly person an experience of the 'right now'.
- Before dying, the last sense is hearing. Before that, touch.
- Stimulate the sense of touch with a client who no longer can react to other sensual stimuli.
- 15 min of touching gently releases oxytocin in the body, which is the 'peace and quiet hormone'.
- 'You can't go wrong by touching the hands.'

Other types of touching were discussed as well; what type of touching is considered as appropriate in care work and what is not. The teacher noted that particularly demented older men in care homes often try and touch the female nurses in inappropriate ways, and that the nurses should then just tell them to stop.

At the end of the day, I took a taxi to the railway station. The taxi driver seemed a nice old chap, and we chatted about the town as a place to live. I was only a visitor, and didn't really know the town that well. He told me that in most respects it is a really nice place to live otherwise—'but there are far too many "refugees", and they do all these crimes: rape women, have gang fights'. I did not—could not—respond to this, not wanting to provoke an argument as I worried the man was the type of racist with whom you just don't argue. I did ask him though, why is it that there are 'so many refugees' in the town. 'Well, it is these reception centres, they receive refugees, who then get settled here.'

Typing the notes on the train back home, I thought about the Filipino students I worked with, the professional registered nurses who vigorously study practical nursing in a foreign language, so as to qualify for work in this country that yet remains foreign to them. And how, maybe many years in the future, the racist taxi driver might well be taken care of by one of these Filipino nurses, overqualified for the job, yet wanting to migrate to Finland and build a life here—welcomed to Finland because the Nation and its ageing population needs the working and caring bodies of the tender loving Filipinos; and because their own Nation needs them to migrate, heroically. Will the racist taxi driver perceive the Filipino nurse as one of the too many 'refugees' in the area? Will he perhaps try to touch a beautiful Filipina nurse in an inappropriate way? Or, perhaps, he cannot himself touch anymore—but only reacts to the touch of the Other. (Tiina's field notes, 2 February 2012)

Through this lengthy extract from Tiina's field diary, we would like to end this chapter by bringing the discussion from the governmental inscriptions of the body in space to the materially concrete relations, movements, resistances and connections that take place within and through the body. In corporeal choreographies of resistance, motion fills space between places in ways such that the trajectory of the movement is never exhaustively prescribed. As an affective potential, choreography articulates the body's capacity to communicate and integrate with other bodies, even as a mere body, a living/dying organism. Here, the possibility, and indeed the necessity of the body to improvise in relation to the other, opens up inevitable and eternal possibilities to create relationalities where no such action should even have been possible. What happens then, if this motion that fills space between places is not imagined to occur between two human bodies as presumably separate entities, but rather between places within the body, provoked by the touch and connection with the body of an Other?

Imagine, the racist taxi driver as a Fragile Old Man. He is living through his last hours before death, in a bed in an old people's home; in his hometown that would be a nice place to live, were there not all those 'refugees'. He is unconscious by now, in palliative care that may or may not be organized adequately in this particular care facility. His body has almost given up its vital functions, preparing to die. The practical nurse on the duty is perhaps a Filipino migrant nurse, whose skin colour or accented speech would cause this racist to label her or him as a 'refugee' or a 'migrant', was he still able to label things using words. But he is not. As a result of his loss of speech, the racist white body has lost his superiority over all others. In the evening, the nurse on duty has a few minutes at the bedside of the Fragile Old Racist. Her hand moves towards the unconscious body. In performing this motion, the colour of the skin or an accented speech suddenly makes no difference. For once, let us imagine, the nurse is not in a hurry. She sits by the bedside for a good while, over fifteen minutes, caressing the hand of the dying. The hand, the touch—these mundane motions of care—provokes a choreographic physical catalyst within the body that no longer can articulate its racist mind to the world. As a human organism not yet dead, the racist body starts to release oxytocin and, within the body, this hormonal motion fills space between the right places, giving the racist the feeling of peace and quiet. Regardless of his racist opinions, the organs of the racist's body somehow miraculously know what to do after his skin (of whatever colour) has sensed fifteen minutes of gentle touching. And the nurse knows this too—whether a plain white Finn, a 'migrant' who's had to 'jump in the sea' to get here, an 'expat' Westerner, or a *bagong bayani* who has to be defined as a hero in order to be as close to plain whiteness as potentially possible. At the event of touch, the body of the racist fails to recognize these differentiations and hierarchies, welcoming and accepting the touch of the other. At the touch, the body of the dying racist thereby resists the touching hand's historicized subjugation. At the touch and when in need of an embodied other, the body resists the superiority of the subjective mind it inhabits.

Vicki Kirby (1997, 61) cites Derrida (1980, 9) to argue that '"the most elementary processes of the living cell" are also a "writing" and one whose "system" is never closed'. This means that the body is 'a shifting scene of inscription that both writes and is written—a scenario where the subject takes itself as its own object'. Thus, as a biological organism, the body is never simply reduced to a docile surface of inscription on which the postcolonial relations and histories are imprinted. To the extent that it is a living body, it is always already a matter of 'peristaltic movements of the viscera, the mitosis of cells, the electrical activity that plays across a synapse, the itinerary of a virus', that is, 'a "text" and a "writing" whose tissue includes all the oozings and pulsings that literally and figuratively make up the differential stuff of the body's extra-ordinary circuitry' (Kirby 1997, 76; see also Fishel 2015; Irni 2016; Preciado 2013). And nevertheless, as a scene of writing, the body also remains as its 'own historical and cultural context' (Kirby 1997, 62), carrying within itself entire trajectories of the past, and thereby enacting new histories, new cultures, new politics, at each and every event bodies come together in corporeal choreographies of encounter.

In the above event then, taking place between and within bodies in a space of care for the elderly, it becomes evident that the relationships between racially differentiated bodies are never exhaustively prescribed. The nurse who is giving care to the dying body of the racist, for instance, may still be a highly educated and experienced registered nurse from a 'third country', whose 'third world' qualifications count for nothing in Finland, or the EU. She may well hold the hand of a dying racist in a care home simply because this is the only job available for a 'migrant'. The silent and almost invisible choreographies of resistance within the body organism are barely enough to dismantle the postcolonial orders where the black and brown bodies not only tend towards whiteness, but are also meant to provide for the corporeal needs of the white across the world. *These postcolonial material-discursive orders of racism and racialization do not disappear at the event of touch and connection that take place between and within the two bodies*. Yet, we argue, it is important to recognize how the body as *a material body organism ultimately refuses to comply with the historically inscribed divisions*. Such recognitions might well be events through which to articulate the world differently. Indeed, as we show in this book, what bodies do as material organisms that are dependent on one another, at the very event of each touch and encounter, overrules and complicates some of the prescribed orders in which the bodies have ended up relating with one another. At each encounter, choreographies and possibilities of the political become improvised in relation to the other, corporeally, over and over again. The way bodies and their mobility enforce encounter *matters*.

NOTES

1. The research in the Philippines was conducted together with Margarita Sakilayan-Latvala. The research has been funded by Tiina Vaittinen's personal grant from the Eino Jutikkala Fund (2013–2016) and by the Academy of Finland project #132403 (2011–2013).

2. In its official statistics of OFW deployment, the Philippines state distinguishes between land-based and sea-based professions. In the figures cited in the text, those deployed in the shipping industry are not included.

3. Practical nurse in Finland refers to an occupational group that can work in several sectors from childcare to the care of the disabled to elder care. In care facilities for the elderly as well as in hospitals, practical nurses are responsible for the manual care work, whereas registered nurses have more clinical training and responsibilities.

Chapter 4

Liminal Space of Relationality and Disturbance

This chapter shifts the focus from the global political economy and racialized categorization of mobile bodies to undocumented migration and asylum seeking in Europe. The chapter engages with (re)thinking embodiment and political agency. It builds on field insights from Finland and France, passing by Greece with the most recent context of the 'migration crisis' of 2015 onwards. It is argued that migrants' bodies and their movement, referred to as choreography, challenge the body politic of the state and the policies of supranational units such as the EU, and thus give new meanings to political agency and enact the event of resistance in multiple ways. Some of these enactments are connected to intentional acts of protest while others are extremely intimate practices. We will contextualize this claim through empirical examples where bodily choreographies intertwine with national, regional and supranational units—a connection that was already suggested and elaborated upon in chapters 1 and 3. What these events through which resistance becomes expressed have in common is their interactional quality, materiality and relationality, as already indicated in the previous chapter and which will be more elaborated in chapter 5: they always concern more than one body. Furthermore, we argue that to fully understand the manifold forms asylum seekers or undocumented migrants corporeally enact resistance, we need to look again at the mundane and minuscule ways in which their agentive capacities are manifested in everyday life.

As we have already explained, in 2015 the phenomenon of asylum seeking became topical across the European continent and also increasingly visible to ordinary citizens. The movement of asylum seekers landing on the shores of Europe and across the continent resulted in many countries in mobilization of racist and right-wing ideologies, but also in acts of solidarity where people mobilized against confrontational discourses. In many locations, people on the move together with locals and transnational activists engaged in visible forms of protest enabled by the large number of people concerned, as for example the case of Budapest Keleti railway station attests (Kallius, Monterescu and Rajaram 2016). To a great extent, the public debate around the question of asylum seeking has been characterized by two extremes. It is our conscious aim in this chapter to make visible the shades of grey between two poles and to illustrate that when looked through the body, it is increasingly difficult to separate between or even recognize governance from resistance, as they both become deeply ambiguous by nature. In line with the argument of the book and other empirical case studies, we discuss the body's agentive capacity with regard to three elements that all are, in their own ways, mundane, intimate, material and rather controversial. First, the force of the mobile body and its capacity for protest also in a state of liminality. Second, the relational politics of desire and sexuality—or the connections between sexuality, desire and politics—that the phenomenon of seeking asylum arises. Third, the way in which material surroundings affect what kinds of acts can be fathomed as manifesting political agency. As explained in chapter 1, our ethnographic work with asylum seekers in Finland (Eeva) and undocumented migrants in France (Anitta) lays the ground for the discussion in this chapter. We recognize the artificialness of the distinction between 'asylum seekers' and 'undocumented migrants' in our respective fields: rather than denoting a definitive characteristic or 'identity' of the concerned person, this distinction refers to the administrative status of the interlocutor at a given time. Moreover, in certain contexts, such as in Calais, having an administrative status as an asylum seeker does not necessarily mean any concrete change as regards to access to accommodation, for example. Also, as we have argued earlier, people racialized as others and considered as unwanted are increasingly mapped into a same category regardless the administrative status.

In framing this chapter under the spatial referent of the camp, we follow our interlocutors' use of the notion of 'camp' to refer to two different types of accommodation arrangements: temporary settlements in the form of makeshift campsites, squatted houses and abandoned industrial buildings in the context of Calais, and reception centres for asylum seekers in Finland. As other authors have eloquently shown, these different kinds of residence arrangements form nodes during the mobile trajectories (Laacher 2002; Yaghmaian 2006; Khosravi 2010; Rygiel 2012; Sigona 2015), and they also re-shape those cities and villages in which they are formed (Agier 2013). In Calais and its surrounding area these arrangements are organized by the migrants themselves and supported by local, national and inter/transnational solidarity actors and organizations (see, e.g., CFDA 2008; EMHRN 2011; Rygiel 2011; Millner 2011). Their existence is not buoyed up by the state or local municipalities, and they are frequently raided and destroyed by different police units. In the Finnish context, on the other hand, the centres are maintained either by municipalities, a selected number of private companies or the Finnish Red Cross with a regular staff and code of conduct. The Finnish case, thus, reflects an institutionalized arrangement that has been set up to accommodate—and administer—those seeking asylum.

CAMP DWELLERS' AGENTIVE CAPACITY

Political agency is conventionally assumed as taking place within the juridical entity of a state, pertaining to the notion of citizenship and manifested most often in terms of electoral rights. Undocumented migrants and asylum seekers cannot claim this sort of political agency: by definition as illegal entrants, they step outside the domain of state-defined legality. Not all states permit the legally residing migrants to take part in local elections, and even those that currently do so, have extended this scope very recently. Yet, we suggest that migrants who aim at transcending national boundaries, especially the state-defined boundaries of (il)legality, engage in and, indeed, enact various forms of political agency (cf. Rygiel 2011, 2012; McNevin 2013; Sigona 2015). This political agency is corporeal as it seeks to contest the national script written on the human body. It is performed, for example, by the very act of taking the route, and by managing to find a livelihood in spaces where one's presence is not legitimated by the correct identity documents or resident status. Mobile individuals engage in corporeal forms of resistance challenging the prevailing order: many are those who traverse deserts, forests and mountains by lorries or on foot, take over-crowded boats in order to cross the Mediterranean Sea and squeeze into lorries to traverse the EU. In so doing, their bodily practice contests the international territorial politics of borders. Once the travel halts and a point has been reached where a life is to be built, reality may sometimes hit hard in the aspired destination (e.g., Khosravi 2010). Ahmad (2008, 2011), drawing on his research with Pakistani smugglees in London's informal economy, describes the irregular working conditions as 'living off borrowed time', which could indeed be extended to many undocumented migrants and asylum seekers across Europe and beyond (e.g., Shakshari 2014 on Turkey).

Both seeking asylum and being undocumented represent ambivalent situations where a person's agency is often thought to be suspended or restricted. During the refugee-status determination process, asylum seekers are mainly supposed to wait for the outcome of the process. Many arrangements that pertain to their daily lives, such as living in a reception centre or being detained, restrict the ways in which they can be heard and perceived. With regard to the undocumented, theirs is a different type of severely restricted access to public space and publicly recognized voice. As they are vulnerable to identity checks and, consequently, deportable by the very lack of a valid visa or a residence permit (de Genova 2002), most are either confined to the shadows of the society or seek to act in as orderly manner as possible, in order not to be remarked. This denotes to a condition of remaining hidden from view or as ordinary as possible in a given context, even if this is very difficult, even impossible in certain urban contexts, such as Calais over the years. For us, choreography and an evented understanding of resistance and agency enable conceptualizing both asylum seekers' and undocumented migrants' agentive capacities without being caught in accounts of social and political powerlessness. Although in the context of the camp, agency and resistance are not celebratory or transcendental features that move the body beyond the state's grip, a univocal account of victimity and powerlessness does not support our empirical experience either (on suspended agency see Ngai 2005). Thus, the question of agentive capacity and resistance whether collective or individual is charged with political meaning.

As the examples selected for this chapter illustrate, corporeal politics takes manifold forms. One way to officialize one's stay is to lodge an asylum claim, and this is what many dwellers in different camps aspire to do. In the context of reception centres, they have already done so. However, as the process towards constructing a common EU-wide asylum policy is yet in the making and, thus far, it seems a fairly unrealistic goal to expect a fully coherent system to evolve in practice, many are those who need to travel from one country to another in search of a safe haven. These circuits form complex choreographies between different countries; they also continue within one country even in the case that a basic right to accommodation is guaranteed in practice. Not all narratives dealing with the same phenomenon easily enter into dialogue with one another; as Iain Chambers writes in his book *Mediterranean Crossings*: 'encountering voices, bodies and lives that exist beyond the official accounts supplied by both colonial and postcolonial power, we are drawn into dissonant narratives' (2008, 59). That is where we aspire to take the readers and, in so doing, illustrate the multifaceted nature of both agency and resistance. However, in the end we do draw some common themes and observations that can be made based on the empirical examples presented in this chapter.

CHOREOGRAPHIES OF (IM)MOBILITY

Let us begin with encounters in northern France, in one of the temporary residences that was opened to provide a shelter for the persons on the move. It was not far from the place where a makeshift transit camp had already been formed some years ago, with local activists and village residents organizing help with practicalities such as food, clothing and washing that are difficult to access in a regional context where undocumented migration is being securitized even more than in many other regions in France, due to the close proximity of the UK border (e.g., Vaughan-Williams 2010) and thus the 'Schengenland' (Walters 2002). The transit camp operated by the Red Cross in Sangatte was closed in late 2002 in the quest of halting unwanted forms of migration: it has been argued since that having a permanent structure for organized residence would constitute a welcome sign for people aspiring to reach the UK (Laacher 2002; CFDA 2008). During extremely cold weather, however, temporary shelters are opened for humanitarian reasons with an agreement by the municipal councils, as it has been realized that to prevent unnecessary deaths, persons in transit need protection against the cold weather during the winter months. Below we present an extract that shows one example of how the border crossing is conceived in the materiality of border control, and how these corporeal trajectories are not fully predefined.

'I would never have imagined travelling like this'

The above quote draws from my discussion with a young Eritrean migrant, Laura, whom I met nearby Calais in January 2010. We were sitting on mattresses laid on the floor in a community space used by local associations to host migrants in transit in the area. The space, unlike many of the squats and makeshift campsites where migrants reside while in the region, was cozy and warm, had proper space for cooking with a stove and sink, and proper toilet facilities. Laura had recently arrived from Saudi Arabia, travelling with an elder man, a friend of her family's. Her family had lived in Saudi Arabia for most of her life; now her father's work contract had ended and the family had to leave. Laura's parents chose to go back to Eritrea, but Laura did not wish to. 'There is nothing for me in Eritrea,' she said. Saudi Arabia was not a country where Laura wanted to reside permanently, had she even been entitled to. Her brilliant English came from years of attending an American private school; as a Christian, she was not allowed in the Muslim public schools. We discussed about the world and family lives, then about her plans for the future: she aspired to study information technology in the UK. Without asking I knew the way in which Laura was planning to reach the northern shore of the English Channel. Laura hugged her legs tight to show how she was to sit during the hours spent in the truck. Tears rose to her eyes. 'I would never have imagined travelling like this,' she said, shivering. It was barely possible to hear what she was uttering and her sadness laid heavy in the air. It was a moment of clarity showing how the global constellation of power and the control of mobility affect the two bodies in such a different manner: a moment earlier we had been two young women discussing their lives, and the very same evening one of us would take her back-bag once more, in order to try to, first, secure herself a place in a truck, and then stay immobile, breathing through a plastic bag to keep the carbon dioxide out of the machine used for controlling the border.

> For Laura, there was a fleeting moment of normalcy when she spoke of her plans for the future, the direction her studies would take once she reached a place where settlement was made possible. Yet the following moment she describes the foreseen means of travel to be able to begin to achieve her dream. She had never imagined travelling like that. (Anitta's field notes and research diary)

Trajectories in the world of forced departures may take weeks, months, even years. Laura's is just one example of what could be called, borrowing from Ian Briuzga (2010), welcoming migrants to the 'Absurdistan of Europe'. Her story includes material realities that the extended border area is constituted of—that is, various control points equipped with a diversity of screening measures for detecting the body shape and breath—and the policies these practices rely on (Amoore and Hall 2009; Vaughan-Williams 2010). Yet, despite the manifold technical equipment designed to prevent unauthorized crossings, these trajectories find their ways when the need for continuing the journey is strong enough. Laura's story also illustrates that immobility is a central feature of the migratory process. Thus, migrant trajectories are not only characterized by constant movement, but also by phases of immobility both coerced and by choice (cf. Runyan et al. 2013). It is also central to notice that geographies are uneven in their potential for mobility and that as Europe witnessed—at an accelerating pace during 2015, states are eager to restrict people's movement by erecting fences and walls to curb unwanted forms of mobility (Brown 2010). But, as we will discuss next, there are other types of restrictions and obstacles for mobility besides material barriers and walls.

Managing with extreme hard physical conditions in the cold weather and snow is part of the requirements to be able to continue the journey. It counts among the many corporeal hindrances to travel and the temporary stay. Below, we show the intermingling of different types of hindrances.

> In the evening of January 27 we visited the 'African squat' (squat Paigniez) to talk with some fifty persons of different nationalities (e.g. Ethiopian, Somalian and Sudanese) from the African continent residing there. The housing conditions were far below any standards for human beings. To give some concrete examples: the abandoned industrial hall had been barred by a cement wall to prohibit access to the building, heating and light were provided by six makeshift fireplaces which enabled the inhabitants to survive the night in low winter temperatures and were also used for cooking, a big hole in the middle of the hall going down three meters onto a trash deposit area and used as a toilet for the residents. We however saw tents and blankets provided by volunteers from local associations, and the atmosphere was friendly and relaxed.
>
> Two days later, the afternoon of January 29, while visiting the health centre for migrants we learnt that the police had evacuated the building (squat

> Paigniez) in the early morning of January 29, arrested the persons residing there and confiscated all blankets, tents, personal belongings and even the wood used for cooking and heating. Three hours later almost all the arrested persons had been released but the confiscated items had not been returned to their owners. Late in the evening we visited the squat for a second time and observed a big change: all the tents, blankets and other items were gone and there was lack of wood to use for heating. Moreover, we noticed that the number of residents had fallen dramatically as the squat was not considered as a safe place after the police operation in the morning. The remaining squat residents were afraid and the atmosphere was very tense. 'How are we to survive the night?' they asked, worried over the temperatures dropping several degrees below zero. Within the short time we stayed there during this second visit, there were two persons who accidently fell into the big hole in the middle because no light was available, fortunately without any visible injuries this time. (Extract from a report for which Anitta served as the main author, EMHRN 2011)

As the extract above shows, harassment is the best word to describe the everyday conditions of encounter between the police and the displaced persons in the area of Calais. From these observations and other narratives collected, the recurrent practice by the police to arrest the undocumented persons and confiscate their personal belongings became evident. The arrests usually last for a few hours after which the persons are released, just to find themselves on the street without their personal belongings and the means to keep warm to survive the following nights in freezing temperatures as was the case 29 January 2010. Personal belongings also represent the concrete sign of humanitarian assistance: sleeping bags, tents, blankets and warm winter clothes are provided by French citizens and migrants' rights activists from other countries to challenge the prevailing national order of things (Malkki 1995b) categorizing individuals by their birthplace and different statuses accorded or not by the states where they end up.

This border zone is often characterized as a site of extreme victimhood with its regular clampdowns on squats and 'jungles', the makeshift campsites where people on the move reside, people lined up in queues awaiting their share of the daily food prepared and handed out by local charities, torn apart shoes and clothes that have seen their best days a long time ago. It is also a site where various police departments (border police, gendarmerie, local police) chase the suspected migrants all around the town centre and the harbour area. Yet, Calais emerges also as a site where minuscule and more macro-level forms of resistance take place; practices of resistance that seek to alter the prevailing status quo and to question the practices of migration management both by the states (in this context, France and the UK in particular), the EU, and the local authorities looking for 'solutions to the local migration problem' in the town of Calais and its immediate neighbourhood. Despite the closing of the Sangatte camp at the end of 2002 and the subsequent destructions of squatted houses and ad hoc campsites, the logics remain the same: the constant game of hide and seek, temporary arrests but also every now and then successful journeys towards the UK or elsewhere. Simultaneously, however, micro-practices of resistance play out in the everyday life in Calais as individuals and groups counter, or at least disturb the smooth functioning of administrative and policing power—'the police' in the Rancièrean sense (see chapter 2; also Rancière 1999: 28–35; Rancière 1998).

ACTS OF PROTEST AND PRACTICES OF RESISTANCE

In Calais, increasing numbers of camp dwellers grew frustrated over the summer of 2015 and a group of some 1,500 people decided to try to gain access to the Tunnel by force (Elgot & Wintour 2015).

As the newspaper article illustrates, the authorities are not the only ones who look for solutions to perceived problems. Also migrants are capable of enacting large-scale protest against practices of governance that from their point of view are regarded as unjust and problematic. The kinds of acts of protest that the article takes up, do not necessarily result from long-term planning, but grow out of frustration towards the lack of perceived possibilities and the determination to fight for a future that would secure better life prospects and greater safety. Such mobilizations are also an elemental part of the disruption that migrants enact (e.g., Rygiel 2011; Millner 2011). Hence, to speak of resistance in the context of the camp only as minute practices and acts would be misleading. The event of resistance can be clearly visible and public, but in the remaining part of the chapter we will, however, focus on acts and events that resist simple interpretation and yet seem to call our attention and articulate something crucial about the corporeal life of those people who inhabit different kinds of camps across Europe.

Although life as an undocumented migrant or asylum seeker is often about moving or trying to keep on moving, there are points when movement comes to a halt. One phase when movement ceases begins with the lodging of an asylum claim. The asylum-determination process can take months or even years, and during this time people's residence in a country is momentarily a tad more permanent. However, when people wait for their asylum claims to be processed in reception centres, they also enter the space of strict control and surveillance. The institution-like centres operate according to certain regulations and rules, which also means that the lives of those who reside in those spaces are under constant scrutiny and observation. Indeed, the securitization and governing of asylum entail complex and multifaceted governmental technologies of surveillance, embedded at a technical level and taking place on a day-to-day basis (see Bigo 2005; Huysmans 2006).

The administration of human mobility, argues Vicki Squire (2009, 93), is extended and dispersed through technical practices of national body politics that are applied according to a rationality of deterrence. In the case of reception centres, designed to accommodate asylum seekers whose claims are being processed, mechanisms of surveillance prevail, albeit in a different way than with border crossings. In the previous example surveillance and control functioned externally, whereas in reception centres it works at an internal level, within the state (see also Squire 2009, 154–155). The political consequences are also somewhat different, when movement as such is less limited in the centres, but asylum seekers are rendered almost invisible and inaudible as they inhabit those spaces. This is what the notion of suspended or restricted agency describes. Being an asylum seeker involves a severely restricted sense of having control over one's own life. Even though in principle people living in reception centres can move freely and live independently, that is, take care of their daily needs such as food, clothing and cleaning, in many even minor situations they are required to turn to the staff of the centre. Furthermore, most centres have an electronic surveillance system with cameras keeping track of movement to and from the centre. People accommodated in reception centres are subject to multiple restrictions and surveillance within the centre, and yet they do not totally condescend to the operational logic of the centre (Puumala 2016). Their interventions can disrupt the functioning of power and claim, although not necessarily intentionally, a part in what happens to them as a result of the asylum-determination process or after that process has reached its end. These events through which agency is manifested are by no means easy to witness, as they may possess only cathartic qualities or simply the promise of transcending the experienced effects of the person's position (see also Conlon 2013). They are, however, all the more material.

> In the bus I continue reading Caroline Moorehead's book *Human Cargo* (2006). The chapter is about the Australian policy of mandatory detention and the conditions that asylum seekers live in. It tells about their anxieties and behaviour caused by mental distress and trauma. I read about self-inflicted violence and pain. I think about the couple sewing their lips together in the centre where I conduct my fieldwork, the abuse of medicine, suicide attempts, and Soran burning the back of his hand with a cigarette. When I asked why he had done such a thing, his simple answer was '*because I felt like it*'. Moorehead's book tells about a person who has similarly burnt his hand with a lighted cigarette. His answer to the question 'why' had been 'because I can't feel anything'. (Eeva's research diary)

Notwithstanding their marginal position, asylum seekers can contest, resist and interrupt the working of mechanisms of surveillance through their movements, which from the point of view of the state (authorities) are considered 'irregular'. Therewith, they keep the potential for politics open and represent these spaces as ones in which the struggle between governance and resistance or agentivity culminates. Moving the body in a direction or in a manner not anticipated/allowed by the rules and norms of the reception centre ultimately dislocates the notion of what political agency can be and where it can be found (also Puumala et al. 2011; Puumala 2016). Moreover, it questions the working of governmental practices at the everyday level and shows that the controlling efforts can in the end be contested by what at a first glance seem to be nothing more than minute gestures. However, we cannot lose sight of the fact that the kinds of acts that we exemplified, arise from a simultaneous sense of hope and hopelessness.

Expressions of resistance and agentive capacities in the context of the camp arise from deeply ambiguous effects and a sense of general disorientation. These expressions could be taken to represent what Sianne Ngai (2005) calls 'ugly feelings' with which she refers to affective gaps and illegibilities, dysphoric feelings and emotional negativity that are connected to situations of suspended agency. These feelings are neither cathartic nor have any therapeutic solutions to the problems they condense. Rather, Ngai (2005, 14) claims that the kinds of acts and events that we have taken up are representative of a meta-feeling where the person feels utterly confused, unsettled or bewildered about what one is feeling. These kinds of situations of general disorientation lack a specified object, which makes them ambiguous and contradictory by nature. They are less organized narratively in the sense that they fall outside of the verbal realm and thus they could easily be labelled politically insignificant.

In terms of resistance the approach arises an intriguing question: could the moving body seek to understand something about the very gesture that the movement is, the environment that movement creates and affects or the capacities and limits of the body concerned (cf. Noland 2009, 106)? For asylum seekers or undocumented migrants voicing oneself and getting a response is not simple. Moreover, language might also be incapable of communicating one's feelings and thoughts especially in emotionally overwhelming situations. The body, in such cases, by moving or gesturing can be fathomed to resist, contest or simply *understand* the political dynamics at work. Movement as a category of political agency does not thus signal a conscious effort towards a goal. It represents a political and relational gesture in itself. Such an approach invites to consider movement as a change of rhythm (of thought), when it becomes clear that current politics can never totally foreclose the space of the political and thus the potential for political engagement.

DESIRE AS AN EVENT OF RESISTANCE

When exploring the political agency of those who are situated in the margins of the present political system and society, that is, live under circumstances where their agentive capacities have been restricted or suspended, attention is often drawn at acts of intentional protest and resistance. As we have discussed in the previous section, this is undeniably one important aspect how agency is manifested and takes shape, but at the same time, it is only one way to fathom how the mobile body interacts with others and systems of governance. In addition to conscious and intentional protest, resistance manifests itself also through events, some of which are extremely mundane and intimate. Yet, such events of resistance disturb or disrupt the logic of control that is at work at a particular moment or with respect to certain people. It is to this dimension that we now turn by attending to the politics of sexuality and desire and understanding how they evoke the question of resistance as a material practice and in terms of intimate events. In much of migration scholarship, apart from queer migration studies (Manalansan IV 2006), people on the move are conceived, at best, as laboursome, asexual beings and not as 'willing and able to experience the full range of human emotions the rest of us take for granted' (Ahmad 2009, 310). In our interpretation, we take again a cue from Rancièrean understanding of disruption. By this we intend that our aim is to look at disturbance from the point of view of the prevailing order, '*police*' following Rancière's terminology, to understand how sexuality and desire as deeply intimate and bodily aspects of human existence, are used for political purposes also beyond the asylum-determination process (cf. Nayak 2015; Shakhsari 2014).

Our interpretation is related to and arises from disturbance to extant ways of categorizing people and constructing hierarchic subject positions. Our thought has been deeply challenged by what we have actually witnessed and learnt through our engagement with people's lives and what happens in various campsites. We have also understood that things become much more messy and complex when they are looked at through seemingly singular events which, however, construct and shape relations among people and between people and structures. The topic of sexual assault committed by asylum seekers, especially the happenings on New Year's Eve 2015 in Cologne, Germany, where women were assaulted by gangs at the city's railway station, shocked widely and gave rise to racist talk against asylum-seeking (Muslim) men, who were labelled as barbarians who cannot and will not commit to 'European values', and who are incapable of controlling their sexual desires. This follows the logic by which the only way to address migrants' sexuality is by pathologizing it, and by focusing on the deviant manifestations of sexuality (e.g., Ahmad 2009, 311).

In what follows, we try to think about ways to come to terms with the politics of sexuality and desire in the camp. In so doing, we are not justifying sexual violence or assault of any kind. But, although much less visible, other kinds of practices, acts and events take place in the camps, and they complicate simple readings of desire, sexuality and control. The notions of victims, perpetrators, abusers and the abused become blurred. What comes forth are relations and, at times, especially the events of negotiation that take place and where positions are sliding and somewhat fluid. Again, as the next example from Greece illustrates, hope and hopelessness intertwine in the notions of resistance and agentive capacity that arise from our reading:

> Our taxi speeds away along the road next to the beach. I'm trying to digest what I just witnessed. I visited one of the ad hoc reception facilities opened by the Greek state and run to the extent possible by voluntary groups and civil society organizations. Funds are lacking, severely. This place has been operating for a few months now; yet saying 'operating' is somewhat an exaggeration. In the meantime, several thousand people reside in the premises and more are about to come if the EU-Turkey migration deal passes. No one can tell what happens to these people: can they trust having their asylum claims rightfully assessed? Could they envision a possibility to travel further to another European country? Could they be deported to Turkey or further away to the countries they first left behind? For the residents there are also more immediate, daily and nightly concerns. There is no proper security within the premises, and sexual assaults and rape are not uncommon. (Anitta's field notes and research diary)

The above excerpt highlights an issue that is frequently discussed among civil society stakeholders, yet surprisingly and unfortunately rarely thoroughly addressed. Earlier, we discussed the manifold surveillance mechanisms in many formal reception facilities. However, on many occasions, simple improvements for ensuring personal safety of the residents, such as locked bedrooms and separate, lockable toilets and bathrooms are not taken into account (cf. Freedman 2007). In the context of ad hoc reception facilities such as the one described above, and makeshift campsites where even the basic facilities are severely lacking, the only way to try to render them safer for the residents is via self-organization of the camp dwellers. Both reception centres and makeshift campsites represent spaces where migration politics and policies intertwine with people's personal and intimate lives. This odd characteristic of these sites actually turns politics into personal and personal into political in ways that escape full, ruptureless descriptions. Agency that in another condition would be considered most intimate and personal can thus become an event, an agentive expression of resistance, even though it might not be outright intended as such. We will concretize our approach with an example before moving further.

> *Disruptive desire?*
>
> An acute issue has emerged in the reception centre. The main topic of the daily staff meeting is the situation in room A188. Six Iraqi men share the room and one of them, Saidi, owns a sex toy, known as the Barbara doll. According to his

roommates he attends to the doll actively, which they, with no private sleeping arrangements, curtains or walls separating their beds from one another, find disturbing. At first there's quite a bit of amusement in the air. Someone wonders where he keeps the doll, another one how he keeps it clean and a third one where he bought it. Discussing such an intimate topic seems to cause the staff to feel a bit awkward and amused. Then a voice says that the man's choice is clearly better than going out picking up women in bars, stalking them in the street or raping somebody. Perhaps, another voice joins in the conversation, it's because he has a wife and twin girls back in his home country. Perhaps he wishes to stay faithful and does not really want just anybody, although he still has his sexual needs. 'But what should *we* do about this?', asks a third person. 'It's clear that the others are bothered and when they tried to confront him on this, Saidi got terribly upset. He assaulted another man and smashed his roommate's cd player. That was not the first time even that he has caused problems'. After these deliberations, the staff agree that an intervention must be expeditiously enacted. The doll, however, seems to make everything more complicated. The staff conclude that transferring this 'troublemaker' to another centre would constitute the best solution, but first they need to have a discussion about the house rules with Saidi.

The next time that I visit the centre I meet Saidi, now known as the 'Barbara doll guy' among the staff. He walks down the corridor and passes the office door carrying a half-full garbage bag. He is wearing a dark blue football shirt, sweatpants and flip-flops. I wonder whether Saidi has already finished 'packing' his things into that garbage bag, or whether he has some other reason to be walking in the corridors with his things. I notice also wondering, when taking a glimpse at the size of the bag, whether the root cause of the whole situation, Barbara, is already there. Saidi mentions that he is ready to leave soon, just a couple of things he needs to pack and say his goodbyes to some friends before leaving. He will be transferred to Rovaniemi, in northern Finland. The centre is more spacious there, at least for the moment, and he might even get a private room. (Eeva's field notes and research diary)

The above vignette illustrates that resistance to practices of surveillance and control is not the only means through which a sense of agency can be born. In an ordinary context, Saidi's behaviour would hardly constitute an example of political agency. However, the reception centre as a certain type of space exposes the asylum seekers to practices that intervene in the most private and intimate aspects of their lives. In that place, having one's sexual needs fulfilled constitutes an act that needs to be subjected to disciplinary measures. The act disrupted, although it most likely did not intend to do so, the order of the reception centre.

As a matter of fact, during their travel and wait undocumented migrants and asylum seekers develop various kinds of relations with others. Sometimes these relations are sexual, and most often the relationships occur between people who are in very different (political) positions with regard to the state. Indeed, many men—asylum seekers in Finland are mostly adult men, especially those who are accommodated in the reception centres (unaccompanied minors being mostly hosted in special facilities)—told about their relationships with Finnish women. Some of those affairs were purely sexual, but some had developed into more 'serious' relationships. It was not, however, the case that only the men were active in looking for company, but also Finnish women. Some of these women were, perhaps, looking for an adventure, and some—who might have stable lives with a husband and children—looked for a change or an adventure.

It is not at all clear, in contrast to the most common story presented in the media, that it is the asylum-seeking men who try to lure women into a relationship and marriage in the hope of a residence permit. Some of the men among whom Eeva has collected her research material told about the women holding the upper hand, them being in the position to determine the way the relationship is going. We begin to understand that better when we realize that on some occasions sex is a means of exchange. Sometimes women can offer the men a place to stay in return for a relationship, even though neither of the parties is that 'serious' about the relationship. These kinds of exchange relationships are far from being uncommon elsewhere either; let us now leave the Finnish reception centres for a moment and return to Calais.

Deterritorialized subjectivities enact new forms of inter-subjectivity in different localized settings. The documentary film *London calling: The Northern Jungle* (2008) by Vincent Nguyen and Jean-Sebastien Desbordes illustrates some of the local relations of help and human contact between the migrants expecting to reach the UK and local dwellers of the Northern Littoral, recounting the story of an elderly lady who has weaved her everyday world into that of the migrants. Anitta's observations in the same region support this too, having witnessed some of the warmth vis-à-vis the displaced, entwined with endurance as regards the repression by the state and local authorities, constant police harassment and other inconveniences which long ago became part of everyday life. In his book depicting undocumented journeys, Behzad Yaghmaian describes another form of daily contact in a newly localized setting, offering with a sexual interpretation:

> The men in Calais and the woods around it were on the road for months. Some had left their homes, and loved ones three or four years before arriving at the last border. They lived in all-male enclaves, spoke to men, ate with men, cried with men, and laughed with men. Except for a few who could buy the sexual services of prostitutes, the *migrants lived dreaming of women, longing for female warmth, touch, and love*.
>
> A few fared better than the rest. In their teens or early twenties, some found love in the arms of older French women, some in their sixties. The women had

> kind and motherly looks, gave the men love and attention, tucked them in their beds, and slept with them. The young men had the comfort of a home and all that came with it. Sex was a central part of the agreement. There was no shower or clean bed *for those failing to deliver. This was a strict business deal, with its own rules and codes of conduct.* (Yaghmaian 2006, 309; our emphasis)

Not a romantic relationship between two persons, explains Yaghmaian, but a business deal involving young male bodies exchanging sex for elderly female bodies, who in turn provided the male bodies with home-like living conditions and care as long as they engaged in delivering sex. Another exchange relationship Yaghmaian describes are encounters between local teenage girls and the battle for their attention between the Kurdish and Afghan groups of migrants. In these exchanges, given that the young girls were not in a position to provide any amelioration of living conditions, as was the case with the elder women with apartments or houses of their own, for the migrants residing in makeshift camps, it was the mutual attention and sex that was a matter of negotiation in these encounters.

> *The scent of women, their laughter, their cries, and their female attributes made the Kurds feel human again.* They joked with the girls—rough jokes, physical jokes, jokes without tenderness—touched them, embraced and squeezed them, and some slept with them. They had sex on the rough ground of their sleeping site at night.
>
> 'They take turns, go to the girls one at a time,' Reza said.
>
> The Kurdish big shots controlled the exit routes. *Power gave them the right to control not only the gates but access to the young and badly desired women of the site.* (Yaghmaian 2006, 310; our emphasis)

Again, no romantic relations—no tenderness in the rough jokes, and exchanging of one girl's body and her sexual services from one man to another—were to be involved. For the migrant men, having the local teenage girls around meant physical satisfaction from smelling women and hearing them laugh and cry, and having sexual needs fulfilled by them. Access to the girls was also a matter of contestation between the groups, the strongest of whom won the battle and thus gained full access. The same applies to the Finnish case. In Eeva's data, it is not uncommon to have references to sex or for the men to make indirect, but nonetheless obvious, suggestions of a sexual relationship. Rough language and extremely detailed descriptions of sexual acts with the odours and positions, together with the characterization of the women or their living conditions abound in some of the men's stories.

Perhaps, desire and sexuality are potentially prominent as events of resistance, because they represent something that is shared by both asylum seekers, undocumented migrants and people who are otherwise positioned, including citizens and migrants with a permanent residence status. Furthermore, they represent deeply intimate elements in our lives, which means that they escape a thorough administrative or, using again Rancièrean terminology, 'police' control. They open a different kind of dynamics between bodies variously positioned and as such can be harnessed and evoked either intentionally or unintentionally to disturb apparently separate subject positions and hierarchies. They evoke a relational politics that resists simple interpretations.

BATHING AS A POLITICAL ACT

Let us now return to Calais and the health centre, which was mentioned earlier in this chapter. The centre was financed by the medical charity organization Médecins du Monde and run by local volunteer nurses and doctors. One of the nurses took the time for a brief discussion with us, mentioning in passing how they had to close the bathroom of the building as it began to be used for showering by persons on the move for whom such facilities were desperately lacking.

> *Showering for comfort*
>
> I enter my hotel room in the centre of Calais. It is two o'clock in the morning. The whole town is quiet, yet in the squatted houses there is a constant murmur of people who chose not to try to cross towards the UK that very night. Others have left, maybe only to return in the early hours of the morning after a futile attempt. They may return later in the afternoon after being released from a few hours of police custody in case of having been caught. Yet others may not return and later we might learn they had managed to get to the other side. Again others end up dying in their attempt to cross. My whole body is frozen. My chin does not stop shaking. I tear away my clothes and let the warm water flow over my body. I remember the saying I heard earlier during the day about comfort showers, those taken more often than once a week which is the standard the local actors aim to uphold for the persons on the move. More frequent ones would be impossible to organize in a fair manner, so as to be available for everyone on an equal basis. This is my second shower of the day. I desperately needed it to become warm again. And I left behind people who had no other chance than stay in the cold, in temperatures below zero in the night and with a freezing wind blowing in from the sea. (Extract from Anitta's field notes and research diary)

The scarcity of available public showers has required associative actors and voluntary workers to be creative in inventing solutions to the bodily need to become clean. In Calais proper, public showers at the disposal of persons on the move have been located at the outskirts of the city; in the surrounding towns and villages solidarity actors have invented complex systems of alternating between various locations. In Calais the showers have frequently been broken, either as a result of the excessive need by hundreds of persons to use the less than ten showers available at a time, or by politically motivated groups that seek to thwart any attempt to fulfil the basic needs of those considered as unwanted. In other towns and villages the engagement of solidarity actors is necessary to organize the transportation of persons from the places where they are staying to the public showers. This is being done with private cars to carry only a few individuals at a time, in order to avoid large crowds in the small localities. It is necessary to keep low key in order to be able to continue the solidarity work in which some of the actors have engaged since the arrival of Kosovars in the late 1990s. The mundane act of washing oneself thus becomes politicized as something the bodies labelled as unwanted are not entitled to do (cf. Twigg 2000, 18–19). The outcasts are not only excluded from the political community but they are also kept at the margins of humanity by not providing for the fulfilment of basic needs, such as cleaning oneself. At the same time fulfilling this mundane need becomes political.

Body washing is one of the human rituals: babies are washed shortly after they utter their first cries, children are taught to wash their hands before joining the table, and dead bodies are washed to prepare them for the afterlife. Religious ceremonies involve contact with purifying water: certain body parts are washed before joining in prayer, or baptism is deployed to differentiate the person belonging to a chosen spiritual family. In fact, as Julia Twigg (2000, 4) claims, bathing and washing play an important part in our domestic lives. Furthermore, these acts sustain and express relationships, endorse values and beliefs and provide a sense of 'existential coherence' to people's lives. Thus, bathing and bathing practices can offer insight into social structures and reinforce relations of power and authority even at a political and societal level.

Washing oneself is not an issue only at the makeshift campsites in Calais. Also asylum seekers who have reached a more stable location and been settled temporarily in a reception centre, often referred to questions of personal hygiene, most often in the context of showering:

> Soran says that he thinks too much, but that he doesn't know how to explain his thoughts to me. He tells that if I stayed at the centre for a night I would understand, if I went to take a shower there, I would understand, but not even the staff understand, or they do not think about it. When he gets to the point of taking a shower in the centre, I slip in: 'I don't want to do that'. He looks at me and says: 'I know you don't, but then you would understand'. (Eeva's field notes)
>
> Benaz tells that when he looks outside, he sees the building on the opposite side of the street from the reception centre where similar people to himself are living. They have their own flats where they have their own showers and

bathrooms. And then he thinks about himself and those others living there in that centre. 'Why do we have to share the shower with 15–16 other people? Why can't we have the same life as Finns?' (Eeva's field notes)

Showering even though mostly perceived as a personal part of our lives, has in the context of asylum seeking become a matter of politics as it illustrates the concrete difference of those who live in the campsites or reception centres. Taking a shower, or having one just for yourself, can then become an element that relates to the socio-political structures according to which our political existence is framed.

POLITICAL EXPRESSIVITY OF THE BODY

This chapter has sought to illustrate that no matter how strictly people are governed and how completely their space of agency and movement is limited, the human body is always capable of changing its direction, (s)pace and rhythm. It unfolds constantly, and in this sense it outdoes sovereign politics that seeks to fix it firmly on a grid. Behaviour and movement that happen in wanton disregard of the regulative ambitions and the resulting disruptive effects of that behaviour illustrate a protest against the rationale that has debased the mobile body and views its presence strictly in terms of a political order, which the person herself or himself is not ready to accept. In this light, agency and gestures may arise prior to cognition, from a sense of dislocation that the loss of control over one's life can cause. When put alongside mass protests or social movements, the body's corporeal resistance is perhaps less romantic or heroic. Because of their ambiguity and multiplicity, however, the body's ways of political expression are all the more disturbing for those who witness them. Through a minor defiance, a reach towards or a move beyond the body discloses existence in ways that displace or rupture the spatial frame within which politics is supposed to occur. The gestures and moves of the agentive body force a response, a change in policies that are designed to direct people's possibilities and space of agency.

The empirical examples that have been discussed in this chapter have in common that the body's experiences are outcomes of involvement in diverse, potentially changing activities and institutions (see Solis 2004, 197). Corporeal and concrete moves and relations form a part of gestural politics, but rather than an organized strategy or a coherent narrative of resistance or agency, that politics is perhaps best conceptualized as passages (cf. Soguk 2000, 433). Those passages are movements, which open space and construct it as meaningful, intelligible and experienceable. Furthermore, the empirical data illustrates that vulnerability also is, from another perspective, a part of agency. Based on her ethnographic insight Christina Zarowsky (2004, 198–199) claims that individual suffering and political experience are linked. She states that instead of isolation, a person's suffering actually embeds them deeply in a moral web. Thus, suffering, emotional experience and people's multiple ways of expressing them connect the singular body to broader political, social and economic structures and forces. As their acts and expressions communicate grief, distress, hope and the pain caused by politics, they suggest a critical engagement and relation between variously categorized bodies (also Schick 2011).

Through Rancière's work, it is possible to appreciate and grasp how people can enact their potential for political agency and disrupt the prevailing police order at least momentarily in their daily lives, through various kinds of relationships and mundane events. Rancière himself emphasizes the quest of those who have no part to have a part. That, in our context, would imply that the people whose stories and lives we have addressed in this chapter, wanted to become members of that community which now excludes them. Yet, it is unlikely that this is the whole 'truth' of those bodies, their dissonant narratives and disruptive acts that take place at various campsites across Europe. Yet, it can be said that notwithstanding their marginal condition, undocumented migrants and asylum seekers negotiate the limits set for their bodies and their existence.

The body is a potent medium of flesh and blood, which carnates a person's capacity to interact with others and her or his surroundings (see also Hewitt 1997, 122; Reischer and Koo 2004, 303). As such it bears the kind of ontological dimension that Coole and Frost (2010, 20) envision that concerns bodies' capacity to interact with other bodies through their gestures and behaviour to evoke a response and prompt forms of judgement that are not necessarily passed through conscious awareness. *Undocumented migrants' and asylum seekers' agentive capacities concern boundaries built between bodies and the efforts to cross these boundaries, so as to raise a different political relationality.* Although the kind of emotive, corporeal and gestural body politic that we have sketched here might sometimes be without clear purpose and direction, it illustrates that the struggles and the everyday relations of undocumented migrants and asylum seekers evoke a variety of dispositions, capacities and potentialities that allow for agency.

Chapter 5

Urban Space of Mundane Interaction

'Bodies which have "traces of dubious origin" (Ahmed 2007, 162) do not sink into public spaces but "stand out"', writes Michele Lobo (2014, 719) in her discussion on practices of everyday multiculturalism in Australia. In chapter 3 we, on the other hand, discussed the attempt of the Philippine state to produce its migrating bodies as an example of a nation which actively seeks proper places for bodies with 'traces of dubious origin' in the international order of things and in its hierarchies of bodies. In chapter 4 we examined how bodies with 'dubious behaviour' disrupted the order of the reception centre. In line with these, this chapter turns to everyday encounters and studies how politically orchestrated anti-immigration projects are enacted in and through the mundane flow of everyday events and how they are resisted. We do that through a case study on Roma beggars in Finland as passing discriminatory encounters between those who belong and 'those who never will' has recently become a common phenomenon in Finnish cities. Roma beggars entering Finland mainly from Romania and Bulgaria constitute a group of distracting and disturbing bodies that 'stand out' and do not find 'their proper place' either in the Finnish streetscape or the society. Although the number of Roma beggars in Finland is minimal—newspapers estimated that 150 beggars had entered Helsinki in the summer 2009—compared to, for example, 40,000 Roma living in camps in Italy (Sigona 2005) and a total of 12 million Roma living in Europe (Reinertsen 2014, 489), some concerned citizens and politicians have found it difficult to cope with the situation, calling it a 'Roma problem'. They have argued for the need to stop this uncontrolled movement of migrating bodies. These anti-immigration opinions have gradually gained a prominent position in the media, especially due to the recent 'influx' of refugees to Europe, including Finland.

The migrating Roma in Finland are part of the internal tension within the EU, a tension that has been legally addressed somewhat differently in various member states but nevertheless produces similar marginalizingly lived experiences for the Roma people (Parker and López Catalán 2014). On the one hand, the EU promotes equal rights for the Roma in the East European member states; on the other hand, it pressurizes those countries to keep the Roma within their borders, thereby denying their right to free movement as EU citizens (Nacu 2011, 138). The legislative challenges have been forcefully fuelled by leading Finnish politicians, such as the Mayor of Helsinki Jussi Pajunen, who in the media identified street disorder explicitly with Roma beggars. For the general public, these type of statements created in their minds a strong contrast between the image of the Finnish model welfare state and the problem of categorizing the migrating Roma who were neither tourists, nor working migrants, nor asylum seekers (Roman 2014, 796). Following Lobo's (2014, 722) reading of Ahmed (2004) the Roma bodies are preceded by the politics of borders at the EU 'as well as ongoing experiences of everyday racism that pathologise, victimize, and racialize our bodies'. The game against beggars provides an example of how anti-immigration projects are transferred to the streets and enacted in the very mundane encounters between differently situated, yet relational, bodies. Leaving aside the legislative discourse as well as politicians as the primary actors of politics, we open up the analysis to include actual encounters, still witnessable as recordings, between Roma beggars, xenophobic citizens and random passers-by with the aim of revealing the collaborative and situated production of racist events and the subtle forms of bodily resistances that bring dissensus into existing social orders. It is within these events that the public space is ongoingly and situatedly accomplished as space of governance, and particularly of resistance, that is, space where political existence is created, negotiated and articulated in novel and innovative ways.

As a specific way of projecting racialist thoughts, this chapter looks at one specific amateur videographic project. An arrogant citizen, 'a man of the law', approaches female Roma beggars, in his words 'women short-changed by the mafia', in an effort to make the Roma problem accessible and visible to a larger audience. His interactional and interpretative strategies make certain bodies stand out from the mundane flow of bodies while at the same time hiding the normal, expected and naturalized bodies, in which he includes himself. The videos, both the actual filming as well as their production, are an integral part of his attempt to control and monitor the beggars and ultimately to strengthen the police order in the Rancièrean sense. He seeks to solve the problem he claims the authorities have turned a blind eye to; his aim is to portray Roma beggars as incompetent and passive victims, the lowest of lows in the (global) political economy and residing outside the constitutional state.

These attempts do not always succeed as we shall see. Spontaneous gestures and breaching of conversational rules open up resistance, resulting in unexpected turns in the choreography.

Treating racist videos circulating on the Internet as ethnographic artefacts of interaction reveals also the spatial policies rooted in racial degradation. Streets and pavements are platforms for moving bodies in the urban landscape. On the street, sporadic bodies are modulated into social gatherings and a pace is set for the intermittent bodily movements to develop into meaningful face-to-face interactions as people rush to work, school, home or services. Streets, squares and pavements are not just about transportation but they present public spaces for relating to other bodies, engaging in mobility as well as for negotiating identities. For citizens, urban encounters make their actions accountable to each other (Goffman 1963). These encounters are not always morally unproblematic events and the democratic principles of public space are constantly questioned (Staeheli 2010). In fact, street-level encounters are occasionally exploited in order to show hostility, exclusion and harassment towards certain groups of people (Listerborn 2015). While we are aware of the racist discourse around us, most of us as citizens do not experience these encounters directly unless somehow expressed, brought before our eyes. The racist videos bring racism to us since they make use of and produce spaces through organizing and standardizing practices for the production of categorial vision (Neyland 2006, 601).

In this chapter, hence, we demonstrate the workings of choreographies of everyday racism as well as the interactional, verbal and bodily methods of resisting it. Multimodal interaction analysis is used to tease out the methodic ways (Garfinkel 1967) in which participants organize social interaction in an orderly and publicly recognizable way. Based on the thick description of videoed encounters between the man behind the camera and female beggars being filmed by the man and later made available on the Internet, the sequential organization of interaction, management of emotions and distribution of identities in those situations is offered to depict the strategic ways in which choreographies of harassment are enacted and showcased for the audience.

In the flow of the argumentation of the book, the chapter builds on an apparatus which advances our objectives in four ways in particular. Firstly, we aim to demonstrate, rather than to build on, the notion of how mobile subjects, categorizations of people and feelings of justice are constitutive of and constituting interactional elements in public spaces. Secondly, the chapter shows how emotions, anger and frustration on the one hand, calmness and detachment, on the other hand, carry on the potentialities of interaction from which the political emerges. Thirdly, the empirical data selected to support the analysis highlights the micro-level functioning of corporeal choreographies as a sequential phenomenon. The video clips offer a rigorous view of the two types of bodies (male citizen and women who are cast out from the national order of things) in the framework of mobility within the European Union, and even within the context of the global political economy. These bodies' positions and relationalities allow us to examine how minuscule embodied strategies and methods are employed to maintain and resist the categorizations inscribed on the bodies. We ask how the movement of bodies makes poverty and belonging visible at the same time as it hides them. Finally, in this chapter, we reconsider the role of a single material artefact, the video recording and the videographer especially, in the process of producing provoking events.

ON DOING AND BEING DISTURBED IN URBAN SPACE

In urban space, the city dwellers take large-scale interactional processes and micro-techniques for granted unless they are forced to pay special attention to their conduct and the overarching structures within which they act. The trigger for attention may be, for example, a manifested ideological stance or a momentary act of frustration that demands people to (re)organize action and mobilize their experiences and accompanying emotions in relation to the concurrent activity (McIlvenny 2015). Streets as public places are interesting also from the membership point of view. Citizens treat fellow citizen primarily in the particular role suggested by the environment and the action whenever they are readily available. Service workers are approached in a particular way (Sorjonen and Raevaara 2014). Policemen have different strategies to approach troublemakers accordingly the role they play either as servants or peacekeepers and the task at hand (Jungleib and Matthews 2014). Beggars are treated differently from street performers who, on the other hand, have different methods of making their actions available and spectacular eye-catching to passers-by (Carlin 2014). This is all evident, but nevertheless the roles are enacted in the interaction and they form a variety of choreographies where bodies relate to each other. When the roles are not this evident, the interaction and the roles suggested, given and predescribed suddenly became relevant as the choreographies can make visible something which was earlier not visible.

Samu's case-study material illustrates the complexities of choreographies, agentive capacities and visibilities of relational bodies. In his case-study material, we encounter a Finnish man who walks along the streets of Tikkurila, a suburb of the city of Vantaa in the Helsinki metropolitan area (see figure 5.1[1]). The man has spotted a woman on the pavement, sitting almost still as if ignorant of the surroundings. As he approaches her, they seem to engage in mutual eye contact. In fact, during the last few steps the

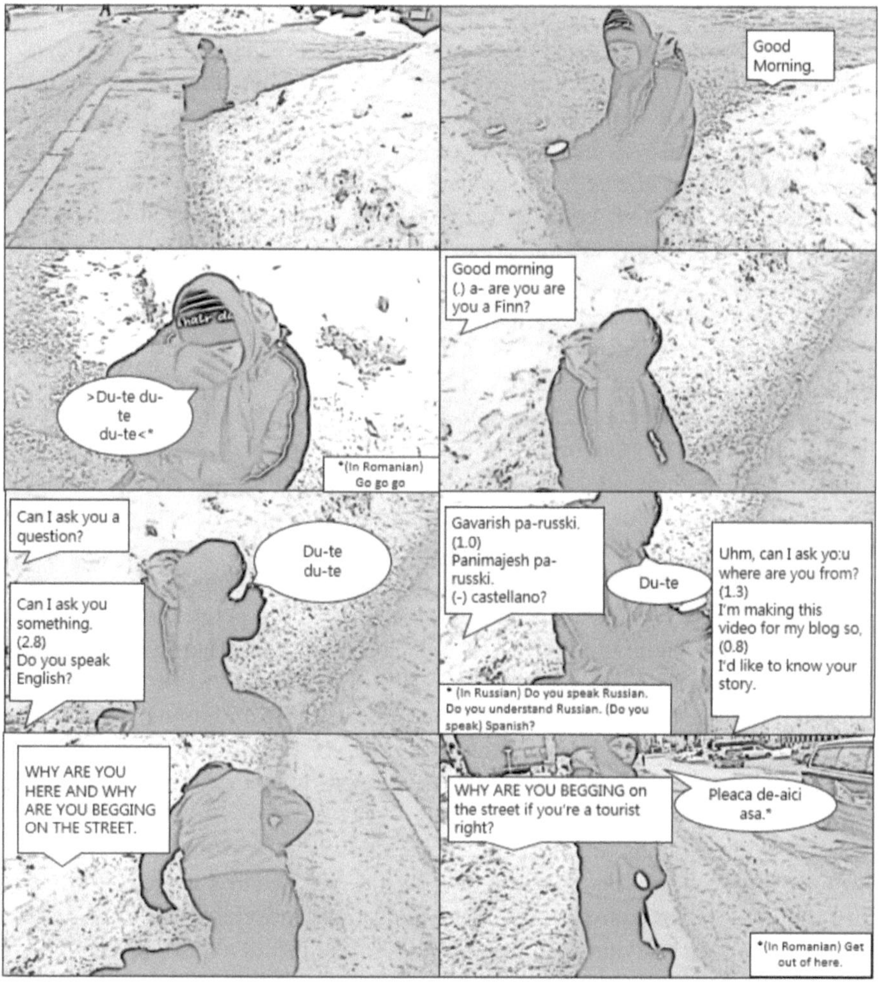

Figure 5.1

man takes, the woman is clearly looking at him with an expectant look on her face. As the man is about to pass her, he says 'good morning', in English, and then stops, facing the woman. On the video recording we experience a lingering pause, see her withdrawing her gaze. She does not greet him back and finally tells the man to go away. 'Du-te, du-te, du-te!', she says in Romanian, reinforcing her speech with a fast series of denial gestures. The man repeats his greeting but, this time instead of waiting for a reply (which never arrives), he stutters a follow-up question: 'A- are you are you a Finn'? The woman withdraws her gaze by turning her head further away from the man. Not just her head but the whole body seems to shun the encounter that just a moment ago seemed to be charged with hopeful expectations for both parties. She repeats the bodily gesture, urging the man to go away. The woman waves her arm several times, and each motion is backed with her uttering yet another 'du-te'. Not sure if the man knows precisely the words the woman says to him but there is no way he could misinterpret her request to leave her alone. What might at first glance have seemed like a passage from an ordinary street encounter between two equal citizens is now irreversibly taken a direction where there is no easy way out for any of the participants, including ourselves as witnesses to the event.

The reason for witnessing this particular street encounter and for making it available for detailed inspection is that the man holds a camcorder. For the participants, this is not a random, unfocused encounter. The man has been intentionally searching out these non-Finnish women 'before which to try out [his] vaunted [self]' (Goffman 1959, 201), engaging in a conversation with them, recording their utterances and movements, editing the data and finally broadcasting the videos on the YouTube and his personal blog. Every effort is made by the man to show these begging Roma women as suspicious characters in a play directed by the EU and silently approved by the Finnish officials whom he blames for allowing endless flocks of Roma people to freely enter Finland and to engage in all kinds of wildcat businesses. Taking to the streets with his camcorder is a reaction—a reaction that seeks to be strategically resistant—to the felt injustice and stress emanating from having to share the public space with beggars. At the same time, his video-projected presence causes stress to the women he harasses as well as to the various audiences both on the street and online.

In his book on behaviour in public places, Erving Goffman (1963, 21) put forward the now-famous notion that it is through situated activity that solitary individuals are transformed into a gathering. Individual experiences are transformed into relevant information about the current situation. Generally, first encounters are packed out with unfinished meaning, projected actions, premises and promises for an interactionally meaningful conversation. This is especially the case in interaction between unknown persons, as has been shown for example in Lorenza Mondada's (2009, 1978) work on strangers asking route instructions. Clearly, there are lots of interactional elements before the first words are exchanged; for example, in the case above the mutual eye contact has made greeting a relevant next action. Spatial disposition and bodily proximity of the participants have to be managed so as to achieve mutual trajectories for interaction and to serve the planned activity between the participants. By approaching the woman sitting on the pavement the man initiates the interaction. His carrying the camcorder makes it a special kind of approach. Possibly because of the fact that the moment will be captured by the camcorder, the woman is keenly following the man as he approaches. She does not simply glance at him, nor is she merely looking at him: she seems to stare at him. To stare at someone articulates a demand for further information. Staring constitutes knowledge production in situ (Garland-Thomson 2009). When the other responds to the stare properly (which may be any kind of open response targeted at the starer), there is a shared interactional space to negotiate the quality and validity of the knowledge.

By turning her head away, by refusing to look at him any longer, the woman indicates the moment of a possible mutual benefit has expired. The man, a newcomer to a situation has transformed her, the solitary individual, and himself into a gathering, and a situated activity (Goffman 1963, 21). She has gathered enough information about the man. She may even anticipate his reasons behind the encounter. She is not willing to go in that direction. She is not ready to give any information about herself. She stands up and walks away.

In his 'general theory of violence as situational process', Randall Collins (2008, 19–20) argues that violent situations are shaped by an emotional field of tension where 'one side appropriates the emotional rhythm as dominator as the other gets caught in it as victim'. Above we witness this as the conversation speeds up into prosecution-like waves of accusations. Making (successful) racist accusations, on one hand, is not a straightforward property but in fact it depends on making other bodies entrained in one's emotions and projects. The beggars' disruptive resistance, on the other hand, resides in breaking the situational rhythm guiding the choreography, showing unexpected agentive capacity and turning the emotional stress towards the offender.

We argue that to explicate these choreographic moves in detail, an interactional approach with pedantic transcription is needed. What the participants themselves make out of it is lamented in and therefore approachable from the interaction captured in the camcorder. The methodological view on interaction followed here is taken from the literature on conversation analysis and multimodal analysis of interaction and mobility (e.g., ten Have 2007; Haddington, Mondada, and Nevile 2013). This literature shows how participants orient towards, use and rely on language, embodied conduct, and spatial and material resources to produce interactional contributions and make sense of the contributions of others. It shows how the institutional roles we tend to take for granted are in fact produced in the interaction. For the man, it is important to develop and fulfil his project-relevant ways of perceiving street space. This way of seeing and showing can be interpreted to be a socially organized activity sustained by and sustaining a community of practice (Neyland 2006, 601). Because the video has been uploaded to YouTube, we as members of the audience become a part of the encounter. As we watch the video and see the corporeal choreographies involved or read the transcription of them, we become witnesses feeling uncomfortable, at points angry or hilarious; we witness a racist attack, an unremarkable encounter or the laying down the law by someone not justified to do so by his profession but by his freedom of speech and feelings about what is just (see Shapiro 2012). What we make out of it is affected by our personal beliefs and motives as much as what we see (see Möller 2013). Where someone sees a racist, someone else sees a hero. Partly for these reasons, it is difficult to witness the event in some other way. Yet, what we can try to see are the interactional means by which participants make the self and others be seen as heroes or racists.

IMPOSING *POLICE ORDER* AND DEONTIC RIGHTS

Despite of the man's first turns in interaction oriented to rather normal and uncontroversial conversation—where a question that does not solicit a response is repeated and often reformulated (Pomerantz 1984b)—and the plea for wanting to know her story for documentary purposes, we can easily start to see where his project is leading to. Annihilating the ethical codes of scientific fieldwork or photographing, the man refuses to respect the woman's demand to leave her alone. He shows no respect of privacy as his questions are penetrating the conversational space like those of an official interrogator casted on a suspect. After several unanswered questions, the woman stands up and prepares to walk out. Walkouts, 'occasions where encounters are brought to a close in an accountably abrupt fashion' (Llewellyn and Butler 2011, 45), are expressions of personal anger and annoyance towards the ongoing event. By walking away, steadily and quietly, the woman makes an effort to determinedly step away from the uncomfortable role she has just been put in and resists the power relations involved in the encounter. Their interactional aims are in conflict: the woman wants to be left alone, the man wants to continue their interaction in order to carry with his project. Yet, they are substantially not disagreeing about anything, at least along the lines of ordinary conversation where disagreements are typically dispreferred, critical assessments softened and challenged in the following turns (e.g., Pomerantz 1984a). They cannot take each other on using words, use those words against the other. The fact that the woman is not responding to the allegations, proving they are false, could of course be a deliberate strategy (Halonen 2005) to simply give the man the cold shoulder. She, however, stops momentarily to tell the man to go away, powered with gestures of denial. By not completely withdrawing, she gives an excuse for the man to continue his sermon. She is trapped in the pace set up by the man. She seems to have agentive capacity the man does not expect her to have, and she refuses the choreography offered to her. By enacting this way, she refuses to become unambiguously and smoothly located within the racialized hierarchy of bodies.

In all subsequent turns the man formulates, he brings in new accusations against the woman and her actions. He also puts into words the labels and classifications he has attached to this woman already before approaching her. For him, she is not just any random citizen taking a short rest in the bustle of the city. She is seen, and introduced to the viewers of the video, as a beggar, someone engaging in the illegal activities. We are witnessing in situ a status-degradation ceremony (Garfinkel 1956) where the man deprives the woman of her former identity and dignity (woman as an individual agent) in order to justify his own alternative categorization (women as ignorant victims of trafficking) and the controlling measures he casts on them. Not just the legality of the women's action but their being in the very cityscape is questioned. Yet, the woman does not fully comply: her resistance is not necessarily individual, intentional and goal oriented, but is something that comes into being relationally as for her the spaces of resistance are by no means readily available.

These are, however, not new questions to her and the people we label as beggars. They may have answered similar questions at the border crossings, when confronted by policemen, security guards, social workers or curious fellow citizens in the street. As Romania and other central European countries have entered the EU and the Schengen agreement, some of the institutional borders have become less demarcating and dominating than opportunities. When entering another EU country as a tourist, there are few restrictions. When it comes to mobility across borders, the Roma women begging and the man following them have equal rights, at least in principle. Although contrary to many European countries such as England and Wales begging is not prohibited in Finland; the Finnish state is following a more general Western tendency in legislation to connect begging with 'deceit, fraud and organized crime' (Adriaenssens and Hendrickx 2011, 24; also Mäkinen 2013) that poses a threat to the public safety. However, as long as there is no reason to believe that the beggar is also taking part in organized crime of some sort, they have a right to be treated as any other citizen in a public space.

The problem has been the partially concealed aim to link the legislation primarily to a certain mobile ethnic group, mainly non-Finnish Roma who enter the country as tourists. The assumption that only Roma beggars constitute a problem has been raised recently, especially in the Finnish news media (Alenius 2012), although these assumptions have internationally and historically much longer traditions (Weber and Bowling 2008). Various suggestions and draft laws have been criticized for ethnic discrimination, for being ineffective or too vague. One option offered has been to not sanction begging itself but the intrusive methods to practice begging (so-called aggressive begging) that may eventually breach public order. In the events analysed here, this would create an awkward situation: the women beg rather quietly (what could be interpreted as a stereotypical passive victim crying for compassion) and they do not physically block other people whereas the man's aggressive filming might well meet the criteria as threatening action in a public space.

The man pretends to know the law and regulations well. However, simply knowing the law does not make him eligible to enforce the law. This disparity between these two situations could be theorized in terms of micropolitics of justice, 'a process in which individuals and collectives are affected by legality/illegalities, such that they participate in a culture of feelings or sensibilities and subsequently engage in discursive encounters about what is just' (Shapiro 2012, 467). For the man and citizens of the same mind, the women begging on the street make visible the complex power relations of mobilities. His action, although targeted at these women, is motivated both by his felt need for a strong police order in the Rancièrean sense and, at the same time, by his distrust of the policies of the government and state officials who allow people like the Roma women to enter Finland without proper grounds. The discursive presence of the police in the conversation crystallizes the motivation of the project the man with the camcorder is trying to manage.

Rights are presented in the event also at the level of interaction. In fact, particular methods are enacted in order to read the law as someone not really eligible to read the law. In the literature on conversation analysis, there has been a recent interest in deontic rights. Someone's deontic authority refers to their right to determine other's future actions. This authority can be acquiesced (deontic congruence) to or resisted (deontic incongruence) in the subsequent turns in the conversation by the other (Stevanovic and Peräkylä 2012). The negotiation is typically stretched over several turns in conversation and the rights are claimed as part of the ongoing project. Figure 5.2 shows an example of the ways in which the rights and claims for authority behind encounters between a group of beggars and the man are negotiated in interaction.

Just before the extract starts insulting words have been exchanged by both parties. Two of the beggars have already prepared to leave the place whereas one woman remains seated. The man gives a sudden exclamation of her identity as a troublemaker and threatens to call the police. The threat is given in two parts within a single turn. First he makes a claim—although grammatically with a conditional clause—about the status of the women as tourists, followed by an authoritative statement about the rights and duties of such a group of people. Based on this account, he threatens to call the police. The threat is repeated in Finnish, this time using the conditional clause, 'if he still finds the women begging on the street'. The deontic right to present the threat emanates from his understanding of the women's status and their activity that is against their performed role as tourists.

Although his threat might be a rhetorical device without any reasonable material consequences in sight, the woman recognizes the single repeated word 'police', thus making the presence of police workable to both parties

Figure 5.2

despite the language barrier. The woman is first looking in the direction of the two other women—away from the man with the camcorder. After the man has uttered the word police twice—first in English and then in Finnish—the woman glances at the man, lifts her right hand up and repeats the word police with a rising intonation as a confirmation question. Although some of the words in the woman's sentence are unclear, her turn with accompanying bodily gestures manifests a challenge to the threat formulated in the man's previous turn. He reasserts his threat by saying that he will really call the police if the women return to the same place. He also extends this threat to include all the women. This manoeuvre is highlighted by the man focusing his camera on the two women who observe the situation from a distance. A sufficient sign of mutual understanding seems to be reached when the woman indicates that the police are not needed. As the man focuses the camcorder again on her, she looks tired and irresolute. The man continues to push further by asking a leading question about the possible outcome the calling of the police might result in for the women. Clearly, the man does not have the authority to decide whether these women could be deported from the country, but seen from the point of view of his interactional project, he can continue harassing as long as his authoritative position remains unchallenged.

RESISTANCE BY REFUSING THE OFFERED CHOREOGRAPHY

The man has basically two motives for undertaking his project: to enlighten and to threaten the women, and in this particular order. Even though the man feels the police is not doing enough to stop beggars, he constantly uses the fear of police to intimidate the women. Yet, both parties seem to share an understanding that the police is not a real threat to their activities. In fact, the quarrel between the man and the beggars is interpretable through the lens of 'coercive care', a desire to safeguard the welfare of 'the vulnerable Other' (Johnsen and Fitzpatrick 2010, 1705); of course, their viewpoints differ in who is considered vulnerable (Finnish society or the women themselves) and in the proper form of care (exclusion from the society or life built on the hospitality of fellow citizens). Challenging the pre-given authoritative positions in conversation often takes place through small devices rather than explicit rebuttals. The episode described in figure 5.2 continues; the women are walking away while the man follows them and continues making threats (see figure 5.3).

In the midst of heated argumentation, accusations are often received with counter-accusations rather than accounted in terms of proving them right or wrong. The next turn is then used to make the other party be seen in an even more sceptic light. The rationale for the man's project of interacting with the Roma women may well be understood by the women. Rather than admitting being engaged in something illegal, the women turn the interactional choreography upside down using other roles and categorizations available within the situation. Again, single words like 'sex' that are understandable despite the lack of common language can be used to resist the pace of degrading accusations. One of the potential reasons for a Finnish man to engage in conversation with a foreign woman is to buy sex. Redefining the situation already defined (Gardner 1995, 217) by building on cultural and social stereotypes is thus a methodic and agentive move available for the women. Now, the threat

Figure 5.3

of calling the police on the public scene is made a potential threat also for the man.

One of the main elements of derogatory choreographies is the emotional stress and dislocation following from the feeling of being out of place. Although the begging body by definition is shown as vulnerable and passive, the enactment of this bodily posture is 'self-selected' by the woman. What the man with his camera tries to do is to take away this agentive capacity, to question its motivation and to expose the faultiness of begging all together. Outrageous actions by the women—running away, trying to grasp the camera or throwing a snowball—stand as proof for the man that he has made a difference by his actions. Achieving this type of racist enjoyment—often discernible in his documentary speech style in the videos—is not, however, always achieved; the women have subtle corporeal methods of resisting being degraded. We have already shown how the insulting turns in conversation can be met with counter-arguments. There are, however, more subtle methods of disturbing racist enjoyment.

The man has spotted once again a young woman, this time sitting in front of the shopping centre's sliding doors. The man—with camcorder in hand—walks straight up to the woman who immediately pulls her jacket up in front of her to cover her head. 'Don't take the picture, leave me!', she shouts. This time he does not waste his time on greetings and polite questions. The man fires up with words (see figure 5.4).

The conversation has quite a different beginning compared to the example in Figure 5.1. Firstly, the participants do not greet each other. Woman's first audible turn is proscription directed to the man who is approaching her with a camcorder. The act is verbal and corporeal as she covers her face with her jacket. Maybe they have already met each other. Maybe this local man has already gained a reputation for harassing beggars and his pursuit is known to the woman. Maybe the Roma women have discussed about this man who is known to approach and film them with little benefits for them. Maybe they have worked out an action plan for the next encounter. Despite all the maybes, what we can say for sure is that the encounter is problematic from the outset. There is no space for reciprocal choreography as overlapping talk and upsetting tone take over from reciprocal search for the common ground. Secondly, the young woman is talkative. They may not speak the same language but they are communicating on an equal footing, occasionally rather effectively. From the conversational flow it is difficult to judge who is in the process of being trapped. Words drop out of her mouth almost as fast as of the man's. Once she unveils her face her hands are subtly gesturing further layers of interpretation. She is not taking part in the choreography offered to her neither is she willing to stay invisible. She challenges the part supposed to be played by her body in that particular place in the process of claiming equality (cf. Davidson and Iveson 2014, 142).

Figure 5.4

By covering her eyes and face—'the portion of the body that socially defines one's public appearance' (Garfinkel 1956, 421)—the woman's body withdraws from the predictable interaction while at the same time remaining visible. The content of her repeated turns channel her insistence not to be photographed or filmed. While the covering of her face might reasonably be an indication of shame or guilt, her dropping of the jacket and revealing her face mark an act of resistance: she makes her, and her kind, agentive capacity visible to the man as her contestation becomes political through 'subjectification' (Davidson and Iveson 2014, 142), that is, this seemingly individual act becomes more than personal. Being exposed to the filming highlights the man's unruly act of filming (cf. Kitzinger and Frith 1999). Her accompanying

Figure 5.5

turn, 'Show me in television brother. Show me', is formulated as a demand that goes along with the given activity. By demanding that the man does better what he is already doing disturbs their power positions as this shifts her body from the place assigned to it and changes the choreography of relational bodies (cf. Rancière 1999, 30). As Celia Kitzinger and Hannah Frith (1999) have remarked about the stories of the sexual harassment of women, these types of efforts to conversationally make the participants equal are often deployed by the less powerful parties. The woman refuses to stay the passive body under suspicion and attack.

After a further round of (unsuccessful) accusations (figure 5.5), the man is trying to save the situation by transforming to his documentary mood: he turns on the camcorder to record the adjacent environment and to capture the physical and social context of the action, as an ethnographer would say. He also switches the language to Finnish with some English words in between. His monologue is interrupted by the woman catching up and repeating the place name Tikkurila right after the man has mentioned it. Repeating it with a slightly different pronunciation, the man nevertheless acknowledges her inserting turn in talk as meaningful. When he also repeats the place name by emphasizing the last syllable of the word, his turn becomes heard as a correction of the woman's pronunciation. He also recognizes the woman as his audience by producing the English language insertion 'center of the fucking universe' in the middle of his turn produced in Finnish. Interrupting a speaker's turn is a blatant form of resistance. The golden rule of conversation—one talks at time—is dismissed not only by interrupting and taking a turn in talk where none is given. Also, the fact that overlapping talk prevails among the participants shows that the limits of friendly interaction have been exceeded.

In a conversation where the more repressed or abjected party leads the discussion, a strategy of resistance can take shape as staged compliance (Gardner 1995, 212). In other words, in the example resistance is expressed through discussing and gesturing in a 'normal' manner—in not being provoked. That is to produce turns where no substantive turn is expected as it challenges the exclusive right of the repressive party to set the agenda and pace of conversation. Rather than trying to get away from and out of the situation or to challenge the man's deontic rights to make accusations, the woman manages the emotional tension (Kitzinger and Frith 1999). The intruding strategy of a one-sided conversation where showers of accusations overshadow the interactional space remains no longer valid and functional. By producing turns in talk, the woman compels the man to follow the logic of dual interaction rather than a blatant telling-off on rights and duties. Although accusing, threatening and roasting never completely cease, the man no longer dictates the autocratic choreography of the encounter. He has to place his accusations in a conversational space shared with another agentive party who is capable of re-routing the flow of conversation, changing the corporeal choreography and, ultimately re-framing the event by producing de-classifications of spatial orders that produce disturbance (cf. Rancière 2001, thesis 7).

MATERIALITY OF THE EVENT AND (IR)RHYTHMIC BODIES

While the long-term project for the man is acknowledged to be to get rid of the Roma beggars, the interactional short-term task is to remove the women from their current spots. With the woman described in the first section, this turned out to be the case. Not only did he succeed in removing, but by following the women for a couple of minutes he also generated a movement applauded by his imaginary audiences; the unwelcome reception of non-citizens, those with no right to be visible and among us. However, the woman in the last case, firmly seated in front of the shopping centre main door, refuses to move and, consequently, becomes invisible again. In fact, while being an active discussant, she never stops her own activity, begging. Seated in front of the main entrance to the shopping centre she also utilizes the movement-regulating space effectively for her own ends. Every time someone passes her, she greets them with a smiling voiced 'moi' (hi, in Finnish). The last greeting seems to be enough for the man. With an annoyed voice, he suddenly says 'moi moi' and stops filming. This leaves us with an impression of the man not completely controlling the situation and his emotional stance. Not succeeding in his project makes him vulnerable in the eyes of people passing by. The man treats these people passing by as outsiders, not belonging to the scene. He purposefully directs his camcorder to face the pavement thus capturing only the legs of those passing by. This might well be an effort to maintain anonymity. Because the focus of filming is on the women, the gazes and postures of those passing-by are not available in the video. He wants to keep outsiders of the events as outsiders unless they are on his side. These are the instances of what Goffman (1963, 24) calls unfocused interaction, 'the kind of communication that occurs when one gleans information about another person present by glancing at him, if only momentarily, as he passes into and the out of one's view'. What is audible, however, are some complaints to the man to leave the women alone. The man does not make these complaints noticeable on the video as they would shatter the image and interpretation that the man wishes later to present to some wider group of citizens.

The situational methods and immediate strategies the participants engage in their encounter are thus embedded in the materiality of making the event. The fact that the encounter has been filmed by the man effects and partly dictates the outcome of the social interaction. Some of the man's turns in the conversation with the beggars are formulated and directed towards a larger audience, the viewers of the video, not for the beggars. This is evident also in the documenting strategies in which the man uses his video recorder to contextualize and to set the scene for the encounter. Within the encounter, there are moments of withdrawal from the ongoing conversation that suggest the instrumental role of the women in the man's derogatory project which seeks to keep the racialized bodies in their proper place. At the interactional level, the man's project leads to situations that are not only threatening in the face of the women but are denying the women any attempt to tactfully 'sign out' from the situation. In this way, the man renders the women's bodies either immobile or moving on his orders. Because '[q]uarrels which seek the humiliation of the opponent through personal invective may achieve degrading on a limited scale' (Garfinkel 1956, 424), the man aims to address the encounter with a larger audience through the act of filming. Although the man engages in a derogatory encounter with these particular women, his invectives are not personal per se. Instead of a private quarrel where the stakes are personal, these are encounters with bodies with different abilities, capabilities, legalities and responsibilities. For the man's project they have to be relational bodies in the sense that his and the audience's bodies relates to them, otherwise they would be of little interest to anyone else. The women, however, try to produce these encounters as intimate—with their turns like 'what have I done to you', 'please leave me alone' and with binding words like 'brother'.

To return to the larger context of mobility and the aim of this book, the sheer existence of Central European beggars in the Finnish cityscapes is an outcome of global politics of mobility and the national and international orders that accompany it. The man of the law seeking to embody the police order in the Rancièrean sense and entering in an unusual conversation with the beggars, engages in the national order of things turning the order into an interbodily relation. He becomes a part of 'body-spaces' that are bound to be always ambiguous, *interzones*, spaces that are not only constraining or conditioning, but also enabling and transformable. The women, breaking away from the preassumed passive role by creating and utilizing the emotional stress and refusing the offered choreographies of bodies, find unexpected spaces to enact their agentive capacity. Our effort to become attuned with the choreographies of the (ir)rhythmic bodies on the video has revealed events that communicate people's understanding of their surroundings, the negotiation of their capacities and the reorganization of relations that exist between people. The encounters between the man and the beggar women took place on the street, a concrete daily environment for people often happily unaware of the offensive acts targeted at the beggars. The chapter has shown that ***no matter how trivial and unprepared these encounters may at first seem, they are events in and through which the choreographies of (ir)rhythmic bodies engage bodies with political existence*.**

NOTES

1. The figures present a combination of a simplified version of Jeffersonian transcription (see Jefferson 2004) with comic storyline based on still images from the actual video. Numbers in round brackets (2.8) measure pauses in seconds. Words in round brackets are indicative of what the speaker may have said although it is not audible in the recording. CAPITALS mark speech that is louder than the surrounding speech. Underscoring indicates stress via amplitude. The talk between >signs< enclose speeded-up talk. Colons (Hi:) show that the speaker has stretched the preceding sound whereas a dash shows a sharp cut-off.

Chapter 6

Conclusions

Unfolding into the Future

William Connolly (2013, 403) has argued that 'appreciation of the fragility of things requires cultivation of greater sensitivity to multiple ways in which contemporary institutions, role definitions and nonhuman processes intersect'. In this book, we have sought to cultivate this sensitivity by studying the complex choreographies of mobile bodies and their sovereign, neoliberal and postcolonial governance. The book as a whole provides tools for imagining the complex entanglements of the choreographies through and in which the human body is not just governed but also capable of changing its prescribed choreography, no matter how strictly people seem to be governed and how fragile their existence is rendered. While our empirical analysis in different chapters represents different phenomena, made meaningful through their particular material-discursive research apparatuses, they are all part of the emergent spatial-temporal entanglements that enact the global at any one moment.

In order to study those institutional, relational and material intersections to which Connolly is referring, we have focused our attention on human mobility. The phenomenon provides us with a point of intersection where different forces, material and nonmaterial, human and nonhuman, converges, where bodies relate to each other in unexpected ways and where they are, simultaneously, normalized, racialized and categorized. People entering their local supermarket may find themselves walking into a heated discussion between a Roma beggar and a man with a camcorder. Later they may be browsing through videos in YouTube only to find that the local transient event had been captured for anti-immigration purposes. They may find out that, at the end of the struggle, there were no clear winners or losers but bodies existing together capable of taking part in manifold choreographies, going along, being 'positioned in' yet disturbing and, indeed, resisting predetermined scripts. All the empirical chapters of the book illustrate how various spaces create room for different kinds of choreographies that may become resistant and manifest agentive capacities that potentially entail political relevance. Thus, our understanding of resistance and agency is not universal or generalizable, but it is embedded in various kinds of potentialities that resonate with the material surroundings and historically inscribed discursive conditions in which bodies relate with one another. Our study of these entanglements shows that the scope of politics embedded in the bodies on the move is greater than often thought.

Through the case studies, this book has brought forth different types of mobile bodies, from asylum seekers to undocumented migrants to labour migrants to beggars, all of whom evoke the question of political space through their encounters with embodied others, both mobile and immobile. The empirical analyses have helped us to rethink notions of agency, resistance and the political beyond the classic categories of participation, representation, different spatial arrangements and their connections. States, as embedded in the global neoliberal political economy, place demands on both the mobility and immobility of bodies, producing highly differential mobilities, and creating hierarchies where bodies are fitted into their 'proper places' in the ever-unfolding national order of things. While examining these differential and differentiated choreographies of resistance we have taken note of the situational and evented ways in which the body is always connected to other bodies. In particular, our inquiry has explored bodies' capacities of evoking the political. Analysing that capacity we have dealt with the inherent tension between political agentiveness on the one hand, and the myriad forms of governance targeted at human bodies on the other.

Human mobility is always a lived and embodied phenomenon that is defined by indefinite material-relational dimensions, all of which are in motion rather than being fixed. Therefore, the treatment of mobility as a simplified, abstracted and generalizable phenomenon is not sufficient. In this book, we have called attention to specific events and concrete temporal-spatial moments, where the political unfolds through with bodies on the move. Our interest in concrete occurrences, specific events and the corporeal elements of mobility has brought us to examine that which is not easily recognizable, visible or articulable. The focus on the evented and relational aspects of existence has enticed us to direct our attention to spaces where the parameters of the fragility of things are enacted and exposed. For us, these events in their actuality represent nodal points that render things vulnerable to one another. Also, they make space for resistances and disruptions that coincide with but do not condescend to the practices of governance, that is, those processes that seek to take hold of the human body (cf. Swyngedouw 2011, 370–376).

ENTANGLED CHOREOGRAPHIES

In this book, we have attended to the idea of choreography, which we define as a situationally enacted interaction framework that provides both a sequential and governing structure for bodily events. We have found the early etymological meaning of choreography (*choreo* deriving from *chora*) useful since it helps us to bear in mind the connection between movement and spatiality. Choreography gears our thinking towards being, as being corporeally in space and inhabiting space, and towards motion that fills space between places. Since choreographies are enacted in corporeal practices, we have focused on both interactional techniques as well as on the actors' absorbing the disciplinary and governing rules. In elaborating the choreographies of corporeal mobility and resistance, we have focused on *events* that emerge in a variety of corporeal-material encounters. By looking closer at embodied, discursive-material encounters and bodily techniques of interaction in our case studies, we show how choreographies are open to surprises and disturbances, where the ordinary may turn into the extraordinary. Examining choreographies of resistance, we have examined the oscillation of bodies as lived, experienced and material configurations, which are simultaneously extremely personal, shared and relational. This approach offers a non-conventional way of conceiving the emergent and constantly reconfigured entanglements of bodies, places, spaces and identities, of matter and discourse and of governance and resistance. In this setting, resistance and agentive capacities are understood in terms of a new materialist ontology that refuses to privilege neither agent nor structure, but rather perceive the two as mutually emergent. Consequently, the politics that the mobile body enacts is a relational politics of becoming that does not succumb to the ideas of sovereign governance that treat mobile bodies as mere objects of governmentality.

The empirical case studies make visible a variety of resistances and agentive practices as they emerge in the intersection of complex temporal-spatial trajectories of governance. Through the case studies we do not, as stated, aim to create a general account of resistance and/or agency in the context of human mobility. Rather, by taking up a selection of differently mobile bodies, we illustrate the abundance of relations that mobile bodies enact and evoke. The choreographies we have examined are thus not generalizable, they cannot be seen as applying to all asylum seekers, undocumented migrants, labour migrants and beggars in Europe at all times. Yet, we find these particular categorizations of mobility illustrative in many ways. They shed light on the differential spatio-temporal choreographies where the practices of governance, including effective state control of boundaries and clear-cut classification of mobile bodies in the national order of things, entangle with the concrete movements of bodies as they improvise and occupy space for their political existence.

Through a closer examination of these groups we have illustrated how the mobile body is itself a factor of turbulence and resistance. The case studies question the ways in which people are categorized in hierarchical positions, through legal, political and economic practices that delineate particular discursive-material spaces of agency. Despite their sometimes extremely disadvantaged positions, mobile bodies possess agentive capacities in the face of both the prevailing global and neoliberal orders and of the national order of things. Mobile bodies articulate their political existence often by improvisation that can be resistant and disturbing.

When enacting political existence, migrants and their forms of mobility give rise to multiple forms of resistance in a variety of locations, simultaneously. Namely, by their very definition of being mobile, bodies in motion never fully fit the routines, flows and rhythms of political-economic orders that are easily perceived as fixed. The orders, however, appear routineous only insofar as bodies move according to the predetermined routine. This is something the very notion of corporeal choreography challenges, for it remains open to surprises each and every time mobile bodies come together and relate with one another. Choreographies of mobility enact dissonance and disruption in orders that appear as ordered, while keeping both orders themselves and the very practices of ordering in constant motion.

We have argued that the body cannot be called resistant as such, and yet it can choreograph events that express and articulate resistance in situations that at first seem redundant of any theorization of politics or agency. Although bodies are moulded by the practices of governance in neoliberal economies, relating with other bodies, they can play with these practices in ways that transform, disrupt, ridicule and even rewrite the scripts of governance. These acts of playfulness are not simply the same thing as collective action taking place outside and against the formal institutions and procedures of the state, and neither is it the same thing as intentionally ridiculing the law through acts of nonconformity. Rather, it is much more subtle, and often unintentional. Yet, it manages to dislocate our conventional ways of thinking about the place and position of the mobile body in the national order of things and beyond.

Our writing falls in the tradition of non-representational theories and we have been interested also in the unintentional nature of corporeal existence. Drawing on a Rancièrean reading of the political, we think that although police (practices of governance) constitutes a society where groups of people are dedicated to specific modes of action and places as well as modes of being, there are events where corporal presence and the choreography of relational bodies disturb the smooth functioning of governance. This disruption and the events of resistance have been the focus of our analysis, and through them we have illustrated how forms of resistance emerge in practice. Our case studies demonstrate how bodies and their politically enacted choreographies are irreversibly entangled with one another, through the neoliberal policies of global capitalism and temporal-spatial heritages of colonialism. The body is a complex site where the discursive entangles with the material, and where temporality and historicity cannot be captured through notions of linearity. As such the body is always multiple: it (re)emerges through various simultaneous material-discursive practices, enactments and configurations. Moreover, the body is—rather than merely representing—many things at the same time and in different contexts, times and places. It is always partially marked and 'already-made-visible', yet with other bodies it is also simultaneously in the process of choreographed becoming.

We have shown how mobile bodies have more agentive capacity and political potential than usually assumed if they are viewed only as objects of, for instance, neoliberal governance, containment, surveillance or bureaucratic practices. Relations of power always entail not only resistance, but also leave space to contest the working of that power. Similarly, bodies on the move remain within the sphere of the *possibly* political because they have agentive capacities beyond the grip of statist governmental and neoliberal (postcolonial) policies. The sphere of the possibly political denotes a politics of becoming where the space for political agency never simply pre-exists, but is (re)articulated through bodies' relationality and choreographies. In this always-already emergent 'intra-action' with other bodies, the political space continues to be (re)configured. Agency is, in this view, about visibility and response-ability, about the possibilities of becoming visible through mutual response (cf. Barad 2007).

WHEN BODIES MOVE, RELATE AND RESIST

As a whole this book has engaged in theorizations of political agency in the context of migration, perhaps most prominently among the acts of citizenship scholarship, and within the field of new materialism (Ahmed 2010a; Coole 2005; Grosz 2010; Häkli and Kallio 2014; Krause 2011; McNevin 2013; Moulin and Nyers 2007; Nyers 2008; Puumala and Kynsilehto 2015; Puumala 2013; Puumala et al. 2011; Redclift 2013; Squire and Darling 2013; Squire 2015; Vaittinen 2014). Although we come close to and share a great deal of common ground in particular with citizenship literature, our position has not been synonymous with it. In our writing, we have at all points surpassed the notion of the sovereign subject—or actually the notion of a singular subject—and emphasized the webs of relations and the political that becomes enacted in those relations. Choreographies of resistance may emerge from a (relational) body, but they never belong to nor completely determined by that body. Thus, we have put forward a relational and evented understanding of politics and resistance that underscores the practices and interactions of bodies whether they are rational or affective, collective or individual, official or every day.

In the empirical chapters we have attended to the meaning of corporeality in the constitution of people's agentive capacities and their choreographies of resistance. The decision to bring together cases to which each of us has attended to in our personal research projects, was made to practice dialectical engagement with empiricism instead of trying to create a ruptureless, representative and uniform theory of either agency or resistance. This choice has enabled us to examine the multiple ways in which the agentive potential and events of resistance take shape and are expressed in different contexts, among various kinds of mobile bodies, their 'others' and networks of power. The cases have underscored an evented understanding of the way state practices and political and economic structures take shape and materialize in migrant's daily lives and their relationships with others.

At the very beginning of our research, it became clear that in addition to a shared theoretical framework, each empirical case required that it should also be analysed through a different philosophical and conceptual discussion. Corporeal choreographies, evented as they are, cannot be analysed by fitting empirical cases to a single rigid theoretical frame. Hence, in the empirical chapters of this book, we have utilized a variety of complementary conceptions of the body—ranging from the body-emitting hormones to the gestural body—and different ways of analysing resistance in its empirical enactments. In a similar fashion, through our empirical analyses we have highlighted the myriad forms of governance and police in the Rancièrean sense. They range from the governance of the global Filipino workforce discursively from a distance to the practices of surveillance and governance in unofficial campsites and reception centres to the self-imposed police order of an ordinary citizen. It is not possible to designate one methodological tool for the study of the relational and material body that moves in space and is in a constant state of becoming and being-in-between. To the extent that analysing corporeal choreographies is about the minute analysis of embodied, material-relational events, no single overarching methodology can be assigned to the study of choreographies. Rather, each methodological apparatus needs to emerge as part of the phenomenon under scrutiny. We have captured these entangled corporeal processes by resorting to ethnographic research and multimodal interaction analysis. Multi-sited ethnography has provided us with a the methodological frame that united chapters 3 and 4, while in order to dig deeper into the details of encounters and the practices of actual interaction, chapter 5 resorted to interaction analysis.

In chapter 3, we showed how the corporeal choreographies enacted in different locations are interwoven with trajectories of the postcolonial, and how the operation of global capitalism continues to divide bodies into categories of differential value—and of differential mobility. The case of the Philippines and its governance of overseas labour from a distance is an example of migration governance crucially different from the European-border surveillance. Yet, the two modes of governance operate in concert when it comes to the governance of mobility through racialization. In Europe, the state power is understood in terms of border controls, capacity for efficient surveillance practices, which aim to keep the unwanted mobile bodies out. In the Philippines, in turn, citizens' mobile bodies are governed so that they choose to migrate across the world in a managed fashion, and here it is helpful for the state to inscribe the Filipinos as better than those who are drowned in the Mediterranean. Depicting the overseas Filipino workers (OFWs) as better than 'the rest', the state intervenes in the postcolonial racialization of its people in the global, inherently racialized discourses of mobility. Here, the police order in the Rancièrean sense administers the visibility of migrant bodies in a particular way and with global implications: it administers postcolonial 'tending towards whiteness' as a globally relevant and shared dynamics of motion, which ties together bodies and their differentially enacted choreographies across space and time. While the Philippine state's intervention is, indeed, non-conformist, yet it is not resistant as it does not enable novel modes of being-in-the-world. The racialized order itself is not much ruptured, or is only minimally shaken, as the Filipino bodies' mobility is rendered desirable only through the reiterated abjection of the undesirable, black others.

In Europe both the bodies of unwanted and undesirable migrants and the more desirable bodies of the OFWs are politically 'ranked' and their productivity is evaluated primarily in economic terms. Humanitarian migrants and Roma beggars are constituted as burdens to the European economy whereas the OFW is brought to the fore as an economically profitable body. However, these economic evaluations, and the differentiated mobilities through which they are enacted, are mutually constructed, as non-white and not-quite-white bodies are made to strive towards the position of whiteness. These (re)inscriptions are not simply about writing bodies but also and in particular about writing bodies differentially, so that the relations between bodies remain hierarchical and racialized. When this setting is interpreted within the apparatus of the postcolonial global political economy, corporeal choreographies become enacted through the writing of difference and power relations in and between bodies; through the (re)writing of certain bodies as potentially present and potentially mobile, while actively abjecting others.

After having examined the wider context of the postcolonial neoliberal economy of governance of mobile bodies, chapter 4 engaged with embodiment and political agency in the context of undocumented migration and asylum seeking in Europe. The chapter evoked the notion of spaces of liminality and the camp as a liminal space within which certain bodies are assigned to wait for the decision whether they can be included in the society or whether they must lead their lives at the margins of society, without a recognized relationship to the state and yet situated within its sovereign borders. For us, both asylum seekers and the undocumented represented examples of suspended agency. It can be argued that their very presence disturbs and, in fact, resists the idea of the national order of things that would comprise a totally orderly organized division of bodies within a given sovereign territory. Furthermore, their liminal positioning makes space for agentive expressions that are disruptive of various manifestations of the political order, and that illustrate the lived materiality of that order and people's acts of protest against it, or events through which they move beyond the order, dislocating its grounding principles. Emotive, corporeal and gestural body politic illustrates that the mundane struggles and relations of undocumented migrants and asylum seekers evoke a variety of dispositions, capacities and potentialities that allow for agency and resistance.

Furthermore, we argued that in the liminal space of the camp, extremely intimate, mundane and private acts and spheres of life become not only illustrative with regard to their agentive capacities, but also deeply politicized. Interpreting such acts, however, always remains without a closure and the relational politics that the camp dwellers enact/evoke is not without contradictions. It can be sporadic, questionable and sometimes seem even irrational, if assessed from the outside. For us, thinking about the politics of sexuality and desire within the camp was a particularly challenging effort. Yet, as the politics of sexuality is, at least in the context of asylum seeking, mostly connected with patronizing or demonizing voices, we wanted to attend to singular events and encounters with people and illustrate the complex processes of subject production and categorization. Desire, sex and sexuality open an ambiguous dynamics between bodies that are categorized and labelled in different ways. When viewed through the analytical prism of choreography, it is possible to claim that they represent events of resistance as they cannot be strictly harnessed to political control. Rather, they evoke a relational politics where positions and power relations are continuously negotiated in ways that resist simple interpretations. In this politics, voice loses its predominance as an expression of agency and the body reveals itself to be a potent medium of flesh and blood that can disrupt practices of governance and the state-centric politics of identification.

While chapter 3 discussed the state-driven attempts to produce its migrating bodies as bodies without 'traces of dubious origin' and chapter 4 focused on those bodies that are often deemed the most undesirable from the state's perspective, the case of Roma beggars respecified the analytical focus to everyday encounters of differential bodies in chapter 5. Members of this particular group of mobile bodies take advantage of the demarcated border-lines within the EU and Schengen Agreement and enter Finland. The focus was on enactments of racism, a particular element within the global politics of mobility.

Through this empirical focus, the chapter illustrated that politically orchestrated anti-immigration projects are enacted in and through the mundane flow of everyday events verbalized and embodied in face-to-face interaction. The Roma beggars certainly carry the marks of 'dubious origin' as they do not tend towards whiteness. Consequently, they are mobile bodies that experience everyday racism and that are pathologized, victimized and racialized as they are nowhere a part of the national order of things. However, also their bodies engage in events where forms of resistance bring dissensus into the existing order of things: racial and derogatory action can be resisted through subtle corporeal choreographies. In this case it was the urban space of streets and pathways where political existence was created, negotiated and articulated in novel and innovative ways through the minutiae of embodied strategies and methods that resist the categorizations inscribed on the bodies. For citizens urban encounters make their actions accountable for each other, but what happens when abjected bodies enact the pre-given urban choreographies anew?

A 'man of the law' who harasses the beggars seeks to make certain bodies stand out from the mundane flow of bodies while at the same time hiding the naturalized, normal, bodies. Through filming his material and technologized presence in the streetscape becomes an integral part of his attempt to control and monitor the beggars, and by doing this he strengthens the police order in the Rancièrean sense. However, the beggars 'speak back', resist, in multiple ways: they challenge the pre-given authoritative positions in conversation; produce emotional stress through bodily gestures and movements; produce disturbances in verbal interaction; turn interactional choreographies upside down using alternative, unexpected, roles and categorizations available within the situation; do not give up their agentive capacity in the situation where it is demanded to be given up; and do not stay invisible. Ultimately, they challenge the part supposed to be played by their bodies in that particular place in the process of claiming equality. By disturbing the power positions through subtle shifts of their bodies in the place assigned to them, they change the choreography of relational bodies. Through the detailed analysis of the choreographies of public encounter, we have cultivated sensitivity to intersections where people communicate their understandings of their surroundings, negotiate their capacities and reorganize relations that exist between people. The encounters between the man and the beggar women took place on the street, in an urban space where seemingly trivial and unprepared encounters appear as events in and through which the choreographies of (ar)rhythmic bodies engage bodies with political existence.

As Alison Mountz (2010, xxiii) has claimed, the putting into operation of sovereign powers through state practice has material and embodied consequences. The perspective put forward by Mountz is one element in theorizing agency and its conditions in the context of migration. However, another and even less theorized question that we have tackled in this book, concerns those relations, be they human, social or political, that form and materialize in spite of state practices. When one looks at the various attempts that were introduced in 2015 or are planned to be put into operation within the EU to control movement and curb unwanted forms of mobility, it could be argued that the possibilities of agency and resistance among migrants are drastically shrinking. Yet, these recent developments illustrate extremely well the processual nature of constructing desirable and unwanted subjects, and the states' as well as the international actors' quest to determine possible relations between subjects by limiting their rights and possibilities.

RELATIONAL COMMUNITY

Throughout this book, our aim has been to problematize the setting which treats some bodies as ontologically and permanently more vulnerable and disadvantaged than others and illustrate the agentive capacities that emerge from, through and in the body. Yet, it cannot be denied that the mobile bodies this book has addressed are situated within the networks of power that affect their capacity to enact themselves and make political claims. Our argument, however, differs from some of the previous research on resistance and space that has looked for those conditions within state structure that provide adequate spaces for resistance and empowerment. References have sometimes been made to marginal spaces (e.g., Routledge 1993) as spaces providing creative resources for empowerment while acknowledging that resistance is ubiquitous (Pile 1997, 16). From the point of view of choreography, we must not give priority to any such theoretical relevance. Rather, looking at the events as choreographed instances of bodies connecting and disconnecting with other bodies, spaces of resistance are affordances that may or may not have been engaged in interaction. To study these passing moments in various choreographies, our diversified methodological tools reveal to us spaces of resistance that are various as well. If a beggar woman has been rudely trapped in a facing formation (or F-formation according to Kendon 1990) with a harassing male citizen, she may use the outsiders walking through their private space of interaction as a space of resistance by simply refusing to leave her spot at the entrance of a mall.

Let the example of the Roma beggar woman take us back to Connolly's (2013) new materialism and the fragility of things that was taken up at the beginning of this chapter. Each of the diverse moments we have analysed embody the characteristics of an event. This is so, because their occurrence is often unexpected or they occur in a place where they were not supposed to do so. Furthermore, their actual happening is relational and as such it can cause disarray in regular and pre-given scripts and roles between bodies and by so doing, disturb the smooth function of the practices of governance. They are amplifications of global processes that make things locally fragile. Yet, the 'global' being directly connected and intra-acting with, and indeed reproduced, reiterated and reformed by the 'local', the minuscule frictions of mundane events are also frictions in the very process of amplification (cf. Vaittinen 2014)—microscopic, passing, barely noticeable disturbances and movement within the global, yet potentially political frictions all the same. These occurrences start with uncertainty, and they foment a wider band of uncertainties as they evolve. Connolly (2013, 404) summarizes events by writing that they 'startle, provoke and energize; they can also disturb, defeat, alienate, overwhelm and sometimes incite resentment against the place of human estate in the cosmos'. In our context of choreographies of mobile bodies, they are complex and even conflictual actualities, that however (re)produce creative responses that engender new results in terms of relationality, visibility and agency.

The pluripotentiality of the events studied in this book, hints at the possibility that the seemingly singular acts are 'creative reverberations back and forth *between* a series of singular acts and collective dispositions that were initially cloudy, in themselves' (Connolly 2013, 405, emphasis in the original). In a way they are strategic moments of uncertainty and creativity as they carry the capacity of being connected while unfolding into the future. Because initially cloudy, the events of resistance can remain invisible to outsiders, especially for those using strong lenses of theoretical vision that take an apparently neutral focus on agency while actually favouring certain spaces, recognizable acts and ontological bodies. Not wishing to remain captive of any singular emancipatory or critical theory, we have 'poached' several discussions and conceptual solutions on the way, trying to stay alert to the forms of resistance and agentive capacities the events analysed could offer.

In this sense, choreographies of resistance are affordances for the participants within their respective mundane social and material environments. James J. Gibson (1979, 9) provides a simple example of affordances as opportunities for action, 'to say that a chair affords sitting may be like saying that a chair with the physical dimensions appropriate to an adult human body expresses a perceivable opportunity for sitting' (Warkentin 2009). Yet, it would be inappropriate to conclude that a chair does not afford other forms of action, nor that sitting could not be afforded by some other materialities. Affordances can be material and social, conforming or distracting; they may be utilized when other capacities for meaning making have been blocked. Analysing events as they unfold within the choreographies of the camp, care and the street has directed our attempt to see resistance in the everyday social life of mobile bodies who are taking part through singular, yet relational acts.

The corporeal choreographies of mobile bodies intersect in multiple, sometimes unforeseen and surprising, ways. When we understand the material-relational body as resisting the hierarchical scripts that the (inter)national order of things inscribes on its surface, we recognize that the migrant's body is not the only body on which the order is being inscribed, for the inscription of bodies is a relational process of differentiation as indicated above. Although this book has mostly focused on the resistances enacted by migrant bodies, it is important to emphasize that our argumentation about differentially mobile bodies does not only abide with those bodies that are labelled as 'migrant', or border-crossing. All bodies' capacity to and affordances of improvisation inevitably open up eternal possibilities to create relationalities where corporeal choreographies are enacted anew. As indicated through the example of a dying racist in need of care in chapter 3, the material body's capacity to communicate and integrate with other bodies does not stop where the subject's capacity to articulate itself stops. As living/dying organisms, our bodies need each other in unexpected ways, and do so in order for politics to remain in motion, and our relatedness open to the political. As a material-discursive entity, the biological body organism that is born from other bodies and continues to depend on others for its mere existence cannot simply be reduced to a docile surface of inscription on which power imposes its rules of being and ordering. We all are mobile bodies, differentially engaged (or differentially having to engage) in the choreographies that our case studies depict. In any event, all of us are part of all this too.

Indeed, the next time we witness a tourist dragging her luggage around the city—an action that might routinely go without notice—we could ask why a Roma beggar walking with her belongings in a plastic bag incites derogatory glares and unfriendly comments while our mobility, even when dragging luggage, does not. We may be confronted by a racist video circulating on social media, where the showing of a young refugee man holding the latest smartphone in one hand and a plastic bag in the other becomes a manifestation of felt injustice in contrast to the indifference experienced when a similar phone is in the hands of the reporter who is a 'proper citizen'. Finally, the whole world has read about the refugee boy who turned a plastic bag into a replica of the football shirt worn by his favourite football player Lionel Messi and where the 'shirt's' subsequent high-media profile and value led to the boy fleeing his home. These are examples of everyday objects forming political subjects, inviting us to see relations and separations between variously positioned bodies. As singular events, their political and resistant agentive capacities may be scarce and go unnoticed even, and especially, by the participants themselves. Yet, as we have argued throughout the book, they challenge our understanding of community and the relational shifting positions of human beings in the choreographies of mobility.

Bibliography

Ackerly, Brooke A., Maria Stern, and Jacqui True, eds. 2006. *Feminist methodologies for International Relations*. Cambridge: Cambridge University Press.

Adey, Peter. 2009. "Facing airport security: affect, biopolitics, and the pre-emptive securitization of the mobile body." *Environment & Planning D: Society and Space* 27: 274–295.

Adriaenssens, Stef, and Jef Hendrickx. 2011. "Street-level informal economic activities: Estimating the yield of begging in Brussels." *Urban Studies* 48: 23–40.

Agier, Michel. 2008. *On the Margins of the World: Refugee Experience Today*. Cambridge: Polity Press.

Agier, Michel. 2013. *Campement urbain: Du refuge naît le ghetto*. Paris: Payot & Rivages.

Ahmad, Ali Nobil. 2008. "Dead men working: time and space in London's ('illegal') migrant economy." *Work, Employment & Society* 22: 301–318.

Ahmad, Ali Nobil. 2009. "Bodies that (Don't) matter: Desire, eroticism and melancholia in Pakistani labour migration." *Mobilities* 4: 309–327.

Ahmad, Ali Nobil. 2011. *Masculinity, Sexuality and Illegal Migration. Human Smuggling from Pakistan to Europe*. Farnham: Ashgate.

Ahmed, Sara. 2004. "Affective economies." *Social Text* 22: 117–139.

Ahmed, Sara. 2006. *Queer Phenomenology: Orientations, Objects, Others*. Durham: Duke University Press.

Ahmed, Sara. 2007. "A phenomenology of whiteness." *Feminist Theory* 8: 149–168.

Ahmed, Sara. 2010a. "Orientations matter." In *New Materialisms: Agency, Ontology, and Politics*, edited by Diana Coole, and Samantha Frost, 234–257. Durham: Duke University Press.

Ahmed, Sara. 2010b. "This other and other others." *Economy and Society* 31: 558–572.

Alenius, Kari. 2012. "Beggars, scammers, discriminated against by the whole of Europe: Romania's Roma in Finnish tabloids, 2008–2011." *Revista Română de*

Studii Baltice si Nordice/The Romanian Journal for Baltic and Nordic Studies 4: 87–109.

Amoore, Louise, and Alexandra Hall. 2009. "Taking people apart: Digitized dissection and the body at the border." *Environment & Planning D: Society and Space* 27: 444–464.

Andersson, Ruben. 2014. *Illegality, Inc: Clandestine Migration and the Business of Bordering Europe*. Oakland: University of California Press.

Appadurai, Arjun. 1997. *Modernity at Large*. Minneapolis: University of Minnesota Press.

Aronsson, Karin. 1998. "Identity-in-interaction and social choreography." *Research on Language and Social Interaction* 31: 75–89.

Badiou, Alain. 2005. *Being and Event*. Translated by Oliver Feltham. London: Continuum.

Barad, Karen. 2007. *Meeting the Universe Halfway: Quantum Physics and the Entanglement of Matter and Meaning*. Durham: Duke University Press.

Barad, Karen. 2012. "Nature's queer performativity." *Women, Gender and Research (Kvinder, Køn og forskning)* 1–2: 25–53.

Bell, Colleen. 2006. "Surveillance strategies and populations at risk: Biopolitical governance in Canada's National Security Politics." *Security Dialogue* 37: 147–165.

Bello, Walden, David Kinley, and Elaine Elinson. 1982. *Development debacle: The World Bank in the Philippines*. San Francisco: Institute for Food and Development Policy.

Bello, Walden, Marissa de Guzman, Mary Lou Malig, and Herbert Docena. 2005. *The Anti-Development State: The Political Economy of Permanent Crisis in the Philippines*. New York: Zed Books.

Bigo, Didier, and Elspeth Guild. 2005. "Policing at a distance: Schengen visa policies." In *Controlling Frontiers: Free Movement into and within Europe*, edited by Didier Bigo and Elspeth Guild, 233–263. London: Ashgate.

Bigo, Didier. 2002. "Security and Immigration: Toward a Critique of Governmentality of Unease." *Alternatives* 27: 62–92.

Bigo, Didier. 2005. "Globalised-in-security: the field and the ban-opticon." In *Translation, Philosophy and Colonial Difference*, edited by John Solomon and Naoki Sakai, 109–157. Hong Kong: Hong Kong University Press.

Bigo, Didier, Sergio Carrera, and Elspeth Guild, eds. 2013. *Foreigners, Refugees Or Minorities?: Rethinking People in the Context of Border Controls and Visas*. Burlington: Ashgate.

Bleiker, Roland. 2000. *Popular Dissent, Human Agency and Global Politics*. Cambridge: Cambridge University Press.

Bleiker, Roland. 2005. *Divided Korea: Toward a Culture of Reconciliation*. Minneapolis: University of Minnesota Press.

Braidotti, Rosi. 2010. "The politics of 'life itself' and new ways of dying." In *New Materialisms: Agency, Ontology, and Politics*, edited by Diana Coole and Samantha Frost, 201–219. Durham: Duke University Press.

Briuzga, Ian. 2010. *Migrerrance: Sur l'exil et les frontières, le cas d'exilés du Pas-de-Calais*. Paris: Migreurop.

Brown, Katrina M. 2012. "Sharing public space across difference: attunement and the contested burdens of choreographing encounter." *Social & Cultural Geography* 13: 801–820.

Brown, Wendy. 2010. *Walled States, Waning Sovereignty*. New York: Zone Books.

Buchan, James. 2006. "Filipino nurses in the UK: A case study in active international recruitment." *Harvard Health Policy Review* 7: 113–120.

Butler, Judith. 1993. *Bodies That Matter: On the Discursive Limits of "Sex."* Routledge: New York.

Butler, Judith. 2004. *Precarious Life: The Powers of Mourning and Violence*. London: Verso.

Campbell, David. 1998. *National Deconstruction: Violence, Identity and Justice in Bosnia*. Minneapolis: University of Minnesota Press.

Carlin, Andrew. 2014. "Working the crowds: Street performances in public spaces." In *City Imagining: Regeneration, Renewal and Decay*, edited by Tara Brabazon, 157–169. Dordrecht: Springer.

Ceyhan, Anna, and Anastasia Tsoukala. 2002. "The Securitization of migration in western societies: Ambivalent discourses and policies." *Alternatives* 27: 21–39.

Chambers, Iain. 2008. *Mediterranean Crossings: The Politics of an Interrupted Modernity*. Durham: Duke University Press.

Choy, Catherine Ceniza. 2003. *Empire of Care: Nursing and Migration in Filipino American History*. Durham: Duke University Press.

Chua, Charmaine. 2015. "Slow Boat to China." *The Disorder of Things Blog*. Accessed 19 April 2016. https://thedisorderofthings.com/tag/slow-boat-to-china/.

Colás, Alejandro. 2002. *International Civil Society: Social Movements in World Politics*. Cambridge: Polity Press.

Coleman, Mat, and Angela Stuesse. 2016. "The Disappearing State and the Quasi-Event of Immigration Control." *Antipode* 48: 524–543.

Collins, Randall. 2008. *Violence: A Micro-Sociological Theory*. Princeton: Princeton University Press.

Conlon, Deirdre. 2013. "Hungering for freedom: Asylum seekers' hunger strikes—rethinking resistance as counter-conduct." In *Carceral Spaces: Mobility and Agency in Imprisonment and Migrant Detention*, Dominique Moran, Nick Gill, and Deirdre Conlon, 133–148. Farnham: Ashgate.

Connolly, William E. 2010. "Materialities of experience." In *New Materialisms: Agency, Ontology, and Politics*, edited by Diana Coole and Samantha Frost, 178–200. Durham: Duke University Press.

Connolly, William. 2013. "The 'New Materialism' and the fragility of things." *Millennium: Journal of International Studies* 41: 399–412.

Coole, Diana. 2005. "Rethinking agency: A phenomenological approach to embodiment and agentic capacities." *Political Studies* 53: 124–142.

Coole, Diana, and Samantha Frost. 2010. "Introducing the new materialisms." In *New Materialisms: Agency, Ontology, and Politics*, edited by Diana Coole and Samantha Frost, 1–45. Durham: Duke University Press.

Coordination Française pour le Droit d'Asile (CFDA). 2008. "Actualités: La loi 'des jungles.'" http://cfda.rezo.net.

Coyle, Angela. 2007. "Resistance, Regulation and Rights. The Changing Status of Polish Women's Migration and Work in the 'New' Europe." *European Journal of Women's Studies* 14: 37–50.

Cresswell, Tim. 1993. "Mobility as resistance: A geographical reading of Kerouac's 'On the Road.'" *Transactions of the Institute of British Geographers* 18: 249–262.

Cresswell, Tim. 1996. *In Place/Out of Place: Geography, Ideology, and Transgression*. Minneapolis: University of Minnesota Press.

Cresswell, Tim. 2010. "Towards a politics of mobility." *Environment and Planning D: Society and Space* 28: 17–31.

Cuban, Sondra. 2013. *Deskilling migrant women in the global care industry*. Basingstoke: Palgrave Macmillan.

Dahle, Rannveig, and Marie Louise Seeberg. 2013. "'Does She Speak Norwegian?' Ethnic Dimensions of Hierarchy in Norwegian Health Care Work Places." *Nordic Journal of Migration Research* 3: 82–90.

Das, Veena. 1998. "Language and body: Transactions in the construction of pain." In *Social Suffering*, edited by Arthur Kleinman, Veena Das, and Margaret Lock, 67–91. Delhi: Oxford University Press.

Das, Veena. 2008. "Violence, gender, and subjectivity." *Annual Review of Anthropology* 37: 283–299.

Dauphinee, Elizabeth, and Christina Masters, eds. 2006. *The Logics of Biopower and War on Terror*. London: Palgrave.

Davidson, Mark, and Kurt Iveson. 2015. "Recovering the politics of the city: From the 'post-political city' to a 'method of equality' for critical urban geography." *Progress in Human Geography* 39: 543–559.

Davidson, Mark, and Kurt Iveson. 2014. "Occupations, mediations, subjectifications: fabricating politics." *Space and Polity* 18: 137–152.

de Genova, Nicholas P. 2002. "Migrant 'illegality' and deportability in everyday life." *Annual Review of Anthropology* 31: 419–447.

Degnen, Cathrine. 2013. "'Knowing,' absence, and presence: the spatial and temporal depth of relations." *Environment and Planning D: Society and Space* 31: 554–570.

Department of Foreign Affairs. 2014. "DFA recognizes the contribution of Filipino seafarers." Accessed 12 April 2016. http://www.gov.ph/2014/06/25/dfa-recognizes-contribution-of-filipino-seafarers/.

Derrida, Jacques. 1980. *Of Grammatology*. Translated by Gayatri Chakravorty Spivak. Baltimore: Johns Hopkins University Press.

Despret, Vinciane. 2013. "Responding bodies and partial affinities in human-animal worlds." *Theory, Culture & Society* 30: 51–76.

Dikeç, Mustafa. 2005. "Space, politics, and the political." *Environment and Planning D: Society and Space* 23: 171–188.

Dillon, Michael, and Julian Reid. 2006. "Global liberal governance: Biopolitics, security and war." *Millennium: Journal of International Studies* 30: 41–66.

Dillon, Michael, and Luis Lobo-Guerrero. 2008. "Biopolitics of Security in the 21st Century." *Review of International Studies* 34: 265–292.

Dillon, Michael. 2007. "Governing through contingency: The security of biopolitical governance." *Political Geography* 26: 41–47.

Diprose, Rosalyn, Niamh Stephenson, Catherine Mills, Kane Race, and Gay Hawkins. 2008. "Governing the Future: The Paradigm of Prudence in Political Technologies of Risk Management." *Security Dialogue* 39: 267–288.

Dossa, Parin. 2008. "Creating politicized spaces: Afghan immigrant women's stories of migration and displacement." *Affilia: Journal of Women and Social Work* 23: 10–21.

Doty, Roxanne L. 2001. "Desert tracts: statecraft in remote places." *Alternatives: Global, Local, Political* 26: 523–543.

Doty, Roxanne L. 2010. "Autoethnography—making human connections." *Review of International Studies* 36: 1047–1050.

Dugan, Kimberly, and Jo Reger. 2006. "Voice and agency in social movement outcomes." *Qualitative* Sociology 29:467–484.

Edkins, Jenny, and Véronique Pin-Fat. 2005. "Through the wire: Relations of power and relations of violence." *Millennium: Journal of International Studies* 34: 1–24.

Edkins, Jenny, Véronique Pin-Fat, and Michael J. Shapiro, eds. 2004. *Sovereign Lives: Power in Global Politics*. New York: Routledge.

Elbe, Stefan. 2008. "AIDS, security and three concepts of risk." *Security Dialogue* 39: 177–198.

Elgot, Jennifer, and Patrick Wintour. 2015. "Calais: man killed as migrants make 1,500 attempts to enter Eurotunnel site." *The Guardian*, 29 July. http://www.theguardian.com/uk-news/2015/jul/29/calais-one-dead-1500-migrants-storm-eurotunnel-terminal.

Epstein, Charlotte. 2007. "Guilty bodies, productive bodies, destructive bodies, crossing the biometric borders." *International Political Sociology* 1: 149–164.

Euro-Mediterranean Human Rights Network (EMHRN). 2011. *Calais, the violence of the border*. Copenhagen: Euro-Mediterranean Human Rights Network.

Fannin, Maria. 2010. "Professional citizenship and medical nationalism in France." *Environment and Planning D: Society and Space* 28: 217–238.

Fassin, Didier. 2011. "Policing borders, producing boundaries. The governmentality of immigration in dark times." *Annual Review of Anthropology* 40: 213–226.

Fishel, Stefanie. 2015. "Microbes." In *Making Things International 1: Circuits and Motion*, edited by Mark B. Salter, 156–170. Minneapolis: University of Minnesota Press.

Foucault, Michel. 1977. *Discipline and Punish: The Birth of the Prison*. Translated by Alan Sheridan. London: Allen Lane.

Foucault, Michel. 1990. *The History of Sexuality, Volume 1: An Introduction*. Translated by Robert Hurley. New York: Vintage Books.

Foucault, Michel. 2008. *The Birth of Biopolitics: Lectures at the Collège de France, 1978–1979*. New York: Palgrave Macmillan.

François, Polet. ed. 2007. *The State of Resistance: Popular Struggles in the Global South*. London: Zed Books.

Freedman, Jane. 2007. *Gendering the International Asylum and Refugee Debate*. Basingstoke: Palgrave Macmillan.

Gammeltoft-Hansen, Thomas, and Nina Nyberg Sørensen. 2013. *The Migration Industry and the Commercialization of International Migration*. Abingdon: Routledge.

Gardner, Carol Brooks. 1995. *Passing by: Gender and Public Harassment*. Berkeley: University of California Press.

Garfinkel, Harold. 1956. "Conditions of successful degradation ceremonies." *American Journal of Sociology* 61: 420–424.

Garfinkel, Harold. 1967. *Studies in Ethnomethodology*. Englewood Cliffs: Prentice-Hall.

Garland-Thomson, Rosemarie. 2009. *Staring: How We Look*. Oxford: Oxford University Press.

Gibson, James J. 1979. *The Ecological Approach to Visual Perception*. Boston: Houghton Mifflin.

Goffman, Erving. 1959/1990. *The Presentation of Self in Everyday Life*. London: Penguin Books.

Goffman, Erving. 1963/1966. *Behavior in Public Places. Notes on the Social Organization of Gatherings*. New York: The Free Press.

Grosz, Elizabeth. 1997. "Ontology and Equivocation: Derrida's Politics of Sexual Difference." In *Feminist Interpretations of Jacques Derrida*, edited by Nancy Holland, 73–101. University Park: The Pennsylvania State University Press.

Grosz, Elizabeth. 2010. "Feminism, materialism, and freedom." In *New Materialisms: Agency, Ontology, and Politics*, edited by Diana Coole and Samantha Frost, 139–157. Durham: Duke University Press.

Guevarra, Anna Romina. 2010. *Marketing Dreams, Manufacturing Heroes: The Transnational Labor Brokering of Filipino Workers*. New Brunswick: Rutgers University Press.

Guillaume, Xavier, and Jef Huysmans, eds. 2013. *Citizenship and Security: The constitution of Political Being*. Abingdon: Routledge.

Haddington, Pentti, Lorenza Mondada, and Maurice Nevile, eds. 2013. *Interaction and Mobility: Language and the Body in Motion*. Berlin: De Gruyter.

Häkli, Jouni, and Kirsi Pauliina Kallio. 2014. "Subject, action and polis: Theorizing political agency." *Progress in Human Geography* 38: 181–200.

Häkli, Jouni. 2015. "The Border in the pocket: Passport as a mobile boundary object." In *Borderities and the Politics of Contemporary Mobile Borders*, edited by Anne-Laure Amilhat Szary and Frédéric Giraut, 85–99. Basingstoke: Palgrave Macmillan.

Halonen, Mia. 2005. "Mä en sit siihe sanonu mitää—Raportit sanomatta jättämisestä ja evidentiaalinen partikkeli sit(te(n)) keskustelun kertomuksissa." *Virittäjä* 109: 272–298.

Haraway, Donna. 1988. "Situated knowledges: The science question in feminism and the privilege of partial perspective." *Feminist Studies* 14: 575–599.

Harvey, David. 2000. *Spaces of Hope*. Berkeley: University of California Press.

Harvey, David. 2001. *Spaces of Capital: Towards a Critical Geography*. New York: Routledge.

Hewitt, Kim. 1997. *Mutilating the Body: Identity in Blood and Ink*. Bowling Green: Bowling Green State University Popular Press.

Hiemstra, Nancy. 2014. "Performing homeland security within the US immigrant detention system." *Environment and Planning D: Society and Space* 32: 571–588.

Holston, James. 2009. "Insurgent citizenship in an era of global urban peripheries." *City and Society* 21: 245–267.

Huysmans, Jef. 2014. *Security Unbound. Enacting Democratic Limits*. Abingdon: Routledge.

Huysmans, Jef. 2006. *The Politics of Insecurity: Fear, Migration and Asylum in the EU*. London: Routledge.

Irni, Sari. 2016. "Steroid provocations: On the materiality of politics in the history of sex hormones." *Signs* 41: 507–529.

IRIN News. 2016. "EU laws designed to deter refugees." 27 January. Accessed 19 May 2016. http://www.irinnews.org/report/102393/eu-laws-designed-deter-refugees.

Jaworski, Adam. 1993. *The Power of Silence, Social and Pragmatic Perspectives*. Newbury Park: Sage Publications.

Jefferson, Gail. 2004. "Glossary of transcript symbols with an introduction." In *Conversation Analysis: Studies from the First Generation*, edited by Gene H. Lerner, 13–31. Amsterdam: John Benjamins.

Johnsen, Sarah, and Suzanne Fitzpatrick. 2010. "Revanchist sanitisation or coercive care? The use of enforcement to combat begging, street drinking and rough sleeping in England." *Urban Studies* 47: 1703–1723.

Jungleib, Lillian, and Katherine Matthews. 2014. "To protect or serve? An interactional analysis of police enforcement of the sit-lie ordinance." Paper presented at The Twentieth Annual Conference on Language, Interaction, and Culture, UCLA, May 2014.

Kafer, Alison. 2013. *Feminist Queer Crip*. Bloomington: Indiana University Press.

Kallius, Annastiina, Daniel Monterescu, and Prem Kumar Rajaram. 2016. "Immobilizing mobility: Border ethnography, illiberal democracy, and the politics of refugee crisis in Hungary." *American Ethnologist* 43: 25–37.

Keith, Michael, and Steve Pile, eds. 1997. *Geographies of Resistance*. London: Routledge.

Kendon, Adam. 1990. *Conducting Interaction: Patterns of Behavior in Focused Encounters*. Cambridge: Cambridge University Press

Keskinen, Suvi, Sall Tuori, Sari Irni, and Diana Mulinari, eds. 2009. *Complying with colonialism: gender, race and ethnicity in the Nordic region*. Farnham: Ashgate.

Khosravi, Shahram. 2010. *"Illegal" Traveller: An Auto-Ethnography of Borders*. Basingstoke: Palgrave Macmillan.

Killias, Olivia. 2010. "'Illegal' migration as resistance: Legality, morality and coercion in Indonesian domestic worker migration to Malaysia." *Asian Journal of Social Science* 38: 897–914.

Kingma, Mireille. 2005. *Nurses on the Move: Migration and the Global Health Care Economy*. Ithaca: Cornell University Press.

Kingsley, Patrick. 2016. "Greece may have deported asylum seekers by mistake, says UN." *The Guardian,* 5 April. http://www.theguardian.com/world/2016/apr/05/greece-deport-migrants-turkey-united-nations-european-union.

Kirby, Vicki. 1997. *Telling Flesh: The Substance of the Corporeal*. New York: Routledge.

Kitzinger, Celia, and Hannah Frith. 1999. "Just Say No? The Use of Conversation Analysis in Developing a Feminist Perspective on Sexual Refusal." *Discourse & Society* 10: 293–316.

Kofman, Eleonore, and Parvati Raghuram. 2006. "Gender and global labour migrations: Incorporating skilled workers." *Antipode* 38: 282–303.

Krause, Sharon. 2011. "Bodies in action: Corporeal agency and democratic politics." *Political Theory* 39: 299–324.

Kuus, Merje. 2013. "Foreign policy and ethnography: A sceptical intervention." *Geopolitics* 18: 115–131.

Kymäläinen, Päivi, and Ari Lehtinen. 2010. "*Chora* in current geographical thought: places of co-design and re-membering." *Geografiska Annaler* 92: 251–261.

Kynsilehto, Anitta, and Eeva Puumala. 2015. "Persecution as experience and knowledge: the ontological dynamics of asylum interviews." *International Studies Perspectives* 16: 446–462.

Kynsilehto, Anitta. 2011a. "Negotiating intersectionality in highly educated migrant Maghrebi women's life stories." *Environment and Planning A* 43: 1547–1561.

Kynsilehto, Anitta. 2011b. *Politics of Multivocality: Encountering Maghrebi Women in France.* Academic Dissertation, Acta Universitatis Tamperensis; 1644. Tampere: Tampere University Press.

Kynsilehto, Anitta. 2014. "Irregularity as a securitized phenomenon in the hubs of transit migration: Practicing security at the EU's external and internal borders." In *Binaries in Battle: Representations of Division and Conflict*, edited by Noora Kotilainen, Aki Huhtinen, and Marja Vuorinen, 141–163. Newcastle upon Tyne: Cambridge Scholars Publishing.

Laacher, Smaïn. 2002. *Après Sangatte… nouvelles immigrations, nouveaux enjeux.* Paris: La Dispute.

Lafazani, Olga. 2013. "A Border within a border: The migrants' squatter settlement in patras as a heterotopia." *Journal of Borderlands Studies* 28: 1–13.

Laurier, Eric. 2004. "The spectacular showing: Houdini and the wonder of ethnomethodology." *Human Studies* 27: 377–399.

Laurier, Eric. 2015. "YouTube: fragments of a video-tropic atlas." *Area.* doi: 10.1111/area.12157.

Leinonen, Johanna, and Mari Toivanen. 2014. "Researching in/visibility in the nordic context: Theoretical and empirical views." *Nordic Journal of Migration Research* 4: 161–167.

Listerborn, Carina. 2015. "Geographies of the veil: Violent encounters in urban public spaces in Malmö, Sweden." *Social & Cultural Geography* 16: 95–115.

Llewellyn, Nick, and Carly W. Butler. 2011. "Walking out on air." *Research on Language and Social Interaction* 44: 44–64.

Lobo, Michele. 2014. "Everyday multiculturalism: catching the bus in Darwin, Australia." *Social & Cultural Geography* 15: 714–729.

Loftsdóttir, Kristin, and Lars Jensen, eds. 2012. *Whiteness and Post-Colonialism in the Nordic Region: Exceptionalism, Migrant Others and National Identities.* Farnham: Ashgate.

Logan, Christie. 2005. "The space between: Phenomenologies of audience, performer, and place for three performances of *Menopause* and *Desire*." *Text & Performance Quarterly* 25: 282–289.

Lopez, Mario. 2012. "Reconstituting the affective labour of Filipinos as care workers in Japan." *Global Networks* 12: 252–268.

Lorenzo, Fely M. E., Jaime Galvez-Tan, Kriselle Icamina, and Lara Javier. 2007. "Nurse migration from a source country perspective: Philippine country case study." *Health Services Research* 42: 1406–1418. Accessed 6 November 2014. http://www.ncbi.nlm.nih.gov/pmc/articles/PMC1955369/.

Madianou, Mirca. 2012. "Migration and the accentuated ambivalence of motherhood: The role of ICTs in Filipino transnational families." *Global Networks* 12: 277–295.

Mains, Susan P., Mary Gilmartin, Declan Cullen, Robina Mohammad, Divya P. Tolia-Kelly, Parvati Raghuram, and Jamie Winders. 2013. "Commentary: Postcolonial migrations." *Social & Cultural Geography* 14: 131–144.

Mäkinen, Virpi. 2013. "Are there fundamental rights for Roma beggars in Europe?" *Political Theology* 14: 201–218.

Malkki, Liisa. 1992. "National geographic: The rooting of peoples and the territorialization of national identity among scholars and refugees." *Cultural Anthropology* 7: 24–44.

Malkki, Liisa. 1995a. *Purity and Exile: Violence, Memory, and Cosmology among Hutu Refugees in Tanzania.* Chicago: The University of Chicago Press.

Malkki, Liisa. 1995b. "Refugees and exile: From 'Refugee Studies' to the national order of things." *Annual Review of Anthropology* 11: 495–523.

Manalansan IV, Martin F. 2006. "Queer intersections: Sexuality and gender in migration studies." *International Migration Review* 40: 224–249.

Mankell, Henning. 2001. *Tea-Bag*. Stockholm: Leopard Förlag.

Manning, Erin. 2007. *Politics of Touch: Sense, Movement, Sovereignty*. Minneapolis: University of Minnesota Press.

Marchart, Oliver. 2007. *Post-Foundational Political Thought. Political Difference in Nancy, Lefort, Badiou and Laclau.* Edinburgh: Edinburgh University Press.

Marcus, George. 1995. "Ethnography in/of the world system: The emergency of multi sited ethnography." *Annual Review of Anthropology* 24: 95–117.

Marcus, George. 1998. *Ethnography through Thick and Thin.* Princeton: Princeton University Press.

Masselink, Leah E., and Shoou-Yee Daniel Lee. 2010. "Nurses, Inc.: Expansion and commercialization of nursing education in the Philippines." *Social Science & Medicine* 71: 166–172. doi:10.1016/j.socscimed.2009.11.043.

Massey, Doreen. 2000. "Travelling thoughts." In *Without Guarantees: In Honour of Stuart Hall*, edited by Paul Gilroy, Lawrence Grossberg, and Angela McRobbie, 225–232. London: Verso.

Matteo, Perin. 2005. "The sense of freedom: the surprise of being-in-common." *Law and Critique* 16: 315–338.

McIlvenny, Paul. 2015. "The joy of biking together: Sharing everyday experiences of vélomobility." *Mobilities* 10: 55–82.

McKay, Steven C. 2007. "Filipino sea men: Constructing masculinities in an ethnic labour niche." *Journal of Ethnic and Migration Studies* 33: 617–633.

McNevin, Anne. 2013. "Ambivalence and citizenship: Theorizing the political claims of irregular migrants." *Millennium: Journal of International Studies* 41: 182–200.

McRuer, Robert. 2006. *Crip Theory: Cultural Signs of Queerness and Disability.* New York: New York University Press.

Megoran, Nick. 2006. "For ethnography in political geography: Experiencing and re-imagining Ferghana Valley boundary closures." *Political Geography* 25: 622–640.

Mezzadra, Sandro, and Brett Neilson. 2013. *Border as Method, or the Multiplication of Labor*. Durham: Duke University Press.

Millner, Naomi. 2011. "From 'refugee' to 'migrant' in Calais solidarity activism: Re-staging undocumented migration for a future politics of asylum." *Political Geography* 30: 320–326.

Mol, Annamarie. 2002. *The Body Multiple: Ontology in Medical Practice*. Durham: Duke University Press.

Möller, Frank. 2013. *Visual Peace: Images, Spectatorship, and the Politics of Violence*. Houndmills: Palgrave Macmillan.

Mondada, Lorenza. 2009. "Emergent focused interactions in public places: A systematic analysis of the multimodal achievement of a common interactional space." *Journal of Pragmatics* 41: 1977–1997.

Moulin, Carolina, and Peter Nyers. 2007. "'We Live in a Country of UNHCR'—refugee protests and global political society." *International Political Sociology* 2: 356–372.

Mountz, Alison. 2004. "Embodying the Nation-State: Canada's Response to Human Smuggling." *Political Geography* 23: 323–345.

Mountz, Alison, and Jenna M. Loyd. 2014. "Constructing the Mediterranean region: Obscuring violence in the bordering of Europe's migration 'Crises.'" *ACME: An International E-journal for Critical Geographies* 13: 173–195.

Mountz, Alison. 2010. *Seeking Asylum: Human Smuggling and the Bureaucracy at the Border*. Minneapolis: University of Minnesota Press.

Nacu, Alexandra. 2011. "The Politics of Roma Migration: Framing identity struggles among Romanian and Bulgarian Roma in the Paris region." *Journal of Ethnic and Migration Studies* 37: 135–150.

Nguyen, Vincent, and Jean-Sebastien Desbordes. "London calling: 'The Northern Jungle.'" Video. November 2008. Accessed 19 May 2016. http://fortresseurope.blogspot.fi/2005/12/london-calling-northern-jungle-part-i.html (part 1); http://fortresseurope.blogspot.fi/2005/12/london-calling-northern-jungle-part-ii.html (part 2).

Näre, Lena, and Camilla Nordberg. 2015. "Neoliberal postcolonialism in the media: Constructing Filipino nurse subjects in Finland." *European Journal of Cultural Studies*. doi: 10.1177/1367549415585557.

Näre, Lena. 2013a. "Ideal workers or suspects: Employers' politics of recognition and the migrant division of care labour in Finland." *Nordic Journal of Migration Research* 3: 72–81.

Näre, Lena. 2013b. "Migrancy, gender and social class in domestic labour and social care in Italy: An intersectional analysis of demand." *Journal of Ethnic and Migration Studies* 39: 601–623.

Nayak, Meghana. 2015. *Who is worthy of protection? Gender-Based Asylum and US Immigration Politics*. Oxford: Oxford University Press.

Neyland, Daniel. 2006. "The accomplishment of spatial adequacy: analysing CCTV accounts of British town centres." *Environment and Planning D: Society and Space* 24: 559–613.

Ngai, Sianne. 2005. *Ugly Feelings*. Cambridge: Harvard University Press.

Noland, Carrie. 2009. *Agency and Embodiment: Performing Gestures/Producing Culture*. Cambridge: Harvard University Press.

Norris, Andrew. 2000. "Giorgio Agamben and the politics of living dead." *Diacritics* 30: 38–58.

Nyers, Peter. 2006. *Rethinking Refugees: Beyond States of Emergency*. New York: Routledge.

Nyers, Peter. 2008. "In solitary, in solidarity: Detainees, hostages, and contesting the anti-policy of detention." *European Journal of Cultural Studies* 11: 333–349.

Olakivi, Antero. 2013. "'In case you can speak Finnish, there is no problem': Reconstructing problematic identity-positions in migrant care workers' organisational discourse." *Nordic Journal of Migration Research* 3: 91–99.

Onuki, H. 2011. "The global migration of care labour: Filipino workers in Japan." In *Feminist ethics and social policy: Toward a New Global Political Economy of Care*, edited by Rianne Mahon and Fiona Robinson, 60–74. Vancouver: UBC Press.

Orlie, Melissa A. 2010. "Impersonal matter." In *New Materialisms: Agency, Ontology, and Politics*, edited by Diana Coole and Samantha Frost, 116–137. Durham: Duke University Press.

Ortiga Yasmin Y. 2014. "Professional problems: The burden of producing the 'global' Filipino nurse." *Social Science & Medicine* 115: 64–71.

Parikka, Jussi. 2012. "New materialism as media theory: Medianatures and dirty matter." *Communications and Critical/Cultural Studies* 9: 95–100.

Parker, Owen, and Óscar López Catalán. 2014. "Free movement for whom, where, when? Roma EU citizens in France and Spain." *International Political Sociology* 8: 379–395.

Parreñas, Rhacel Salazar. 2001. *Servants of globalization: Women, migration and domestic work*. Stanford: Stanford University Press.

Parreñas, Rhacel Salazar. 2003. "The care crisis in the Philippines: Children and transnational families in the New Global Economy." In *Global woman: Nannies, maids and sex workers in the new economy*, edited by Barbara Ehrenreich and Arlie Russell Hochschild, 39–54. London: Granta.

Parreñas, Rhacel Salazar. 2005. *Children of global migration: Transnational families and gendered woes*. Stanford: Stanford University Press.

Parreñas, Rhacel Salazar. 2008. "Transnational fathering: Gendered conflicts, distant Disciplining and Emotional Gaps." *Journal of Ethnic and Migration Studies* 34: 1057–1072.

Pehkonen, Samu. 2016. "Choreographing the Performer-Audience Interaction." *Journal of Contemporary Ethnography*. doi: 10.1177/0891241616636663.

Penttinen Elina. 2008. *Globalization, prostitution and sex-trafficking: A Corporeal Politics*. London: Routledge.

Perera, Suvedrini. 2013. "Oceanic corpo-graphies, refugee bodies and the making and unmaking of waters." *Feminist Review* 103: 58–79.

Philippine Overseas Employment Authority. 2014. "2010-2014 Overseas Employment Statistics." Accessed 12 April 2016. http://www.poea.gov.ph/stats/2014%20POEA%20Stats.pdf.

Pile, Steve. 1997. "Introduction: Opposition, political identities and spaces of resistance." In *Geographies of Resistance*, edited by Steve Pile and Michael Keith, 1–32. London: Routledge.

Pomerantz, Anita. 1984a. "Agreeing and disagreeing with assessments: Some features of preferred/dispreferred turn shapes." In *Structures of Social Action. Studies in Conversation Analysis*, edited by J. Maxwell Atkinson and John Heritage, 152–163. Cambridge: Cambridge University Press.

Pomerantz, Anita. 1984b. "Pursuing a response." In *Structures of Social Action. Studies in Conversation Analysis*, edited by J. Maxwell Atkinson and John Heritage, 57–101. Cambridge: Cambridge University Press.

Preciado, Paul B. 2013. *Testo Junkie: Sex, Drugs, and Biopolitics in the Pharmacopornographic Era.* New York: Feminist Press.

Pugliese, Joseph. 2002. "Penal asylum: refugees, ethics, hospitality." *Borderslands e-journal* 1(1). Accessed 20 December 2015. http://www.borderlands.net.au/vol1no1_2002/pugliese.html.

Puumala, Eeva, and Samu Pehkonen. 2010. "Corporeal Choreographies: Failed Asylum-Seekers Moving from Body Politics to Bodyspaces." *International Political Sociology* 4: 50–65.

Puumala, Eeva, Tarja Väyrynen, Anitta Kynsilehto, and Samu Pehkonen. 2011. "Events of the body politic: A nancian reading of asylum-seekers' bodily choreographies and resistance." *Body & Society* 17: 83–104.

Puumala, Eeva. 2013. "Political life beyond accommodation and return: Rethinking relations between the political, the international, and the body." *Review of International Studies* 39: 949–968.

Puumala, Eeva, and Anitta Kynsilehto. 2015. "Does the body matter? Determining the right to asylum and the corporeality of political communication." *European Journal of Cultural Studies*, online early view, doi:10.1177/1367549415592898

Puumala, Eeva. 2016. *Asylum Seekers, Sovereignty and the Senses of the International: A Politico-Corporeal Struggle.* London: Routledge.

Raghuram, Parvati. 2004. "The difference that skills make: Gender, family migration and regulated labour markets." *Journal of Ethnic and Migration Studies* 30: 303–321.

Rajaram, Prem Kumar, and Carl Grundy-Warr, eds. 2007. *Borderscapes: Hidden Geographies and Politics at Territory's Edge.* Minneapolis and London: University of Minnesota Press.

Rajaram, Prem Kumar, and Carl Grundy-Warr. 2004. "The Irregular Migrant as Homo Sacer: Migration and Detention in Australia, Malaysia, and Thailand" *International Migration* 42: 33–64.

Rajaram, Prem Kumar. 2013. "Historicising 'asylum' and responsibility." *Citizenship Studies* 17: 681–696.

Rancière, Jacques. 2003. "Politics and aesthetics: an interview." Translated by Forbes Morlock. *Angelaki: Journal of the Theoretical Humanities* 8: 191–211.

Rancière, Jacques. 1998. *Aux bords du politique*. Paris: Gallimard.

Rancière, Jacques. 1998. *Disagreement.* Minneapolis: University of Minnesota Press.

Rancière, Jacques. 1999. *Disagreement. Politics and Philosophy.* Translated by Julie Rose. Minneapolis: University of Minnesota Press.

Rancière, Jacques. 2001. 'Ten theses on Politics.' *Theory and Event*, 5: 3 thesis 7, available at: http://www.after1968.org/app/webroot/uploads/RanciereTHESESONPOLITICS.pdf, accessed 25 May 2016.

Rancière, Jacques. 2006. *The politics of aesthetics*. London: Continuum.
Rancière, Jacques. 2009. "A few remarks on the method of Jacques Rancière." *Parallax* 15: 114–123. doi:10.1080/13534640902982983.
Redclift, Victoria. 2013. "Abjects or agents? Camps, contests and the creation of 'political space.'" *Citizenship Studies* 14: 308–321.
Reinertsen, Anne Beate. 2014. "Writing race Roma 3D: 'Put in vase to give nice,' She said—and nobody's here but me." *Qualitative Inquiry* 20: 489–494.
Reischer, Erica, and Kathryn S. Koo. 2004. "The body beautiful: Symbolism and agency in the social world." *Annual Review of Anthropology* 33: 297–317.
Rodriguez, Robyn Magalit. 2010. *Migrants for Export: How the Philippine State Brokers Labor to the World.* Minneapolis: University of Minnesota Press.
Rogers, Barry. 2011. "Pioneering international recruitment in Finland." In *Managing Mindsets: Applying the Cultural Advantage in your Business,* edited by Mijnd Huijser and Danaë Huijser, 82–88. The Netherlands: Ayn Press.
Roman, Raluca B. 2014. "Trans-national migration and the issue of 'ethnic' solidarity: Finnish Roma elite and Eastern European Roma migrants in Finland." *Ethnicities* 14: 793–810.
Rose, Mitch. 2002. "The seductions of resistance: Power, politics, and a performative style of systems." *Environment and Planning D: Society and Space* 20: 383–400.
Routledge, Paul. 1993. *Terrains of resistance: Nonviolent Social Movements and the Contestation of Place in India.* Westport: Praeger Publishers.
Runyan, Anne Sisson, Amy Lind, Patricia McDermott, and Marianne H. Marchand, eds. 2013. *Feminist (Im)Mobilities in Fortress(ing) North America: Rights, Citizenships, and Identities in Transnational Perspective.* Farnham: Ashgate.
Rygiel, Kim. 2011. "Bordering solidarities: migrant activism and the politics of movement and camps in Calais." *Citizenship Studies* 15: 1–19.
Rygiel, Kim. 2012. "Politicizing camps: Forging transgressive citizenships in and through transit." *Citizenship Studies* 16: 807–825.
Salter, Mark B. 2004. "Passports, mobility, and security: How smart can the border be?" *International Studies Perspective* 5: 91–71.
Salter, Mark B. 2006. "The global visa regime and the political technologies of the international self: Borders, bodies, biopolitics." *Alternatives: Global, Local, Political* 31: 167–189.
Salter, Mark B. 2007. "Governmentalities of an Airport: Heterotopia and Confession." *International Political Sociology* 1: 49–66.
Sampson, Helene. 2003. "Transnational drifters or hyperspace dwellers: an exploration of the lives of Filipino seafarers aboard and ashore." *Ethnic and Racial Studies* 26: 253–277.
Schick, Kate. 2011. "Acting out and working through: Trauma and (in)security." *Review of International Studies* 37: 1837–1855.
Security Dialogue. 2008. "Special issue on security, technologies of risk and the political" 39(2–3).
Shakhsari, Sima. 2014. "The queer time of death: Temporality, geopolitics, and refugee rights." *Sexualities* 17: 998–1015.

Shapiro, Michael. 1997. *Violent Cartographies: Mapping Cultures of War*. Minneapolis: University of Minnesota Press.

Sheller, Mimi, and John Urry. 2006. "The new mobilities paradigm." *Environment and Planning A* 38: 207–226.

Shildrick, M. 2002. *Embodying the Monster: Encounters with the Vulnerable Self*. London: Sage Publications.

Shildrick, M. 2012. *Dangerous Discourses of Disability, Subjectivity and Sexuality*. Palgrave MacMillan.

Sigona, Nando. 2005. "Locating 'The Gypsy Problem'. The Roma in Italy: Stereotyping, labelling and 'Nomad Camps.'" *Journal of Ethnic and Migration Studies* 31: 741–756.

Sigona, Nando. 2015. "Campzenship: Reimagining the camp as a social and political space." *Citizenship Studies* 19: 1–15.

Soguk, Nevzat. 1999. *States and Strangers: Refugees and Displacements of Statecraft*. Minneapolis: University of Minnesota Press.

Soguk, Nevzat. 2000. "Poetics of a world of migrancy: Migratory horizons, passages, and encounters of alterity." *Global Society* 14: 415–442.

Soja, Edward W. 2010. *Seeking Spatial Justice*. Minneapolis: University of Minnesota Press.

Solis, Jocelyn. 2004. "Narrating and counternarrating illegality as an identity." In *Narrative Analysis: Studying the Development of Individuals in Society*, edited by Colin Daiute and Cynthia Lightfoot, 181–199. London: Sage.

Sorjonen, Marja-Leena, and Liisa Raevaara. 2014. "Complexity of routine action." Paper presented at Complexity of (inter)action symposium, University of Oulu, October 2014.

Sparke, Matthew. 2008. "Political geography: Political geographies of globalization III—resistance." *Progress in Human Geography* 32: 423–440.

Spillman, Lyn, and Brian Conway. 2007. "Texts, Bodies, and the Memory of Bloody Sunday." *Symbolic Interaction* 30: 79–103.

Spivak, Gayatri. 1988. "Can the subaltern speak?" In *Marxism and the Interpretation of Culture*, edited by Cary Nelson, and Lawrence Grossberg, 271–313. London: Macmillan.

Squire, Vicki, ed. 2011. *The Contested Politics of Mobility: Borderzones and Irregularity*. London: Routledge.

Squire, Vicki. 2014. "Desert 'trash': posthumanism, border struggles, and humanitarian politics," *Political Geography* 39, 11–21.

Squire, Vicki. 2009. *The Exclusionary Politics of Asylum*. Basingstoke: Palgrave MacMillan.

Squire, Vicki, and Jonathan Darling. 2013. "Enacting rightful presence: Justice and relationality in city of sanctuary." *International Political Sociology* 7: 59–74.

Squire, Vicki. 2015. *Post/humanitarian Border Politics between Mexico and the US: People, Places, Things*. Basingstoke: Palgrave MacMillan.

Staeheli, Lynn A. 2010. "Political geography: Democracy and the disorderly public." *Progress in Human Geography* 34(1): 67–78.

Stevanovic, Melisa, and Anssi Peräkylä. 2012. "Deontic Authority in Interaction: The Right to Announce, Propose, and Decide." *Research on Language and Social Interaction* 45: 297–321.

Swyngedouw, Eric. 2009. "The antimonies of the postpolitical city: In search of a democratic politics of environmental production." *International Journal of Urban and Regional Research* 33: 601–620.

Swyngedouw, Erik. 2011. "Interrogating post-democratization: Reclaiming egalitarian political spaces." *Political Geography* 30: 370–380.

Takhar, Shaminder. 2007. "Expanding the Boundaries of Political Activism." *Contemporary Politics* 13: 123–137.

ten Have, Paul. 2007. *Doing Conversation Analysis: A Practical Guide*. 2nd edition. London: Sage Publications.

Tulbert, Eve, and Marjorie H. Goodwin. 2011. "Choreographies of attention: Multimodality in a routine family activity." In *Embodied Interaction: Language and Body in the Material World*, edited by Jürgen Streeck, Charles Goodwin, and Curtis LeBaron, 79–92. Cambridge: Cambridge University Press.

Tungohan, Ethel. 2013. "Reconceptualizing motherhood, reconceptualizing resistance." *International Feminist Journal of Politics* 15(1), 39–57.

Twigg, Julia. 2000. *Bathing: The Body in Community Care*. London: Routledge.

Tyler, Imogen. 2013. *Revolting subjects: Social abjection and resistance in neoliberal Britain*. London: Zed Books.

Vaittinen, Tiina, Margarita Sakilayan-Latvala, and Päivi Vartiainen. Forthcoming. "Registered nurses for export, caregivers to import? Filipino nurse education as a devalued commodity in Finland/EU." In *Ethics and the Marketization and Internationalization of Higher Education: Buying and Selling in the Global Education Marketplace*, edited by Anna Popovic and Fred Dervin. Palgrave.

Vaittinen, Tiina. 2014. "Reading care chains as migrant trajectories: A theoretical framework for the understanding of structural change." *Women's Studies International Journal* 47: 191–202.

Vaittinen, Tiina. 2015. "The power of the vulnerable body: A new political understanding of care." *International Feminist Journal of Politics* 27: 100–118.

Vaittinen, Tiina. Forthcoming. "The global biopolitical economy of needs: Filipino nurses in ageing Finland." PhD diss., University of Tampere.

Valluy, Jérôme, ed. 2005. "L'Europe des camps. La mise en écart des étrangers." *Cultures et Conflits*, spring 2005. Available at https://conflits.revues.org/1710.

Vaughan-Williams, Nick. 2010. "The UK border security continuum: Virtual biopolitics and the simulation of the sovereign ban." *Environment and Planning D: Society and Space* 28: 1071–1083.

Väyrynen, Tarja. 2016. "Rethinking national temporal orders: the subaltern presence and enactment of the political." *Review of International Studies*. doi:10.1017/S0260210515000595.

Walker, Robert B.J. 1994. "Social movements/world politics." *Millennium—Journal of International Studies* 23: 669–700.

Walters, William. 2002. "Mapping Schengenland: Denaturalizing the border." *Environment and Planning D: Society and Space* 20: 561–580.

Warkentin, Traci. 2009. "Whale agency: Affordances and acts of resistance in captive environments." In *Animals and Agency: An Interdisciplinary Exploration*, edited by Sarah E. McFarland and Ryan Hediger, 23–43. Leiden: Brill.

Weber, Leanne, and Benjamin Bowling. 2008. "Valiant beggars and global vagabonds. Select, eject, immobilize." *Theoretical Criminology* 12: 355–375.

Williams, J., and Massaro, V. 2013. "Feminist geopolitics: Unpacking (in) security, animating social change." *Geopolitics* 18: 751–758.

Woon, Chih Yuan. 2013. "For 'emotional fieldwork' in critical geopolitical research on violence and terrorism." *Political Geography* 33: 31–41.

Wrede, Sirpa, and Lena Näre. 2013. "Glocalising care in the Nordic countries: An Introduction to the special issue." *Nordic Journal of Migration Research* 3: 57–62.

Yaghmaian, Benzad. 2006. *Embracing the Infidel: Stories of Muslim Migrants in the Journey West*. New York: Bantam Dell.

Zaman, Habiba and Gina Tubajon. 2001. "'Globalization from below': Feminization of Migration, Resistance and Empowerment—A Case Study." *Canadian Journal of Development Studies* 22: 1109–1129.

Zarowsky, Christina. 2004. "Writing trauma: emotion, ethnography and the politics of suffering among Somali returnees in Ethiopia." *Culture, Medicine and Psychiatry* 28: 189–209.

Zetter, Roger. 1991. "Labelling refugees: Forming and transforming a bureaucratic identity." *Journal of Refugee Studies* 4: 39–62.

Index

About the Author

The writing team, led by Professor Tarja Väyrynen, consists of five scholars, namely Eeva Puumala, Anitta Kynsilehto, Samu Pehkonen and Tiina Vaittinen from Tampere Peace Research Institute (TAPRI), University of Tampere. The multidisciplinary research team (Political Sciences, International Relations, Peace and Conflict Studies, Human Geography) has expertise in mobility studies, resistance, political theory and corporeal politics. The group has published, for example, in Body & Society, Environment and Planning A, Review of International Studies, International Political Sociology, International Feminist Journal of Politics, and has contracted books with Manchester University Press and Routledge.

www.ingramcontent.com/pod-product-compliance
Ingram Content Group UK Ltd.
Pitfield, Milton Keynes, MK11 3LW, UK
UKHW041626031026
15587UKWH00001B/84

9 781783 486724